Exploding the Myths Surrounding ISO9000

A practical implementation guide

Exploding the Myths Surrounding ISO9000

A practical implementation guide

ANDREW W. NICHOLS

IT Governance Publishing

IT Governance Publishing
IT Governance Limited
Unit 3, Clive Court
Bartholomew's Walk
Cambridgeshire Business Park
Ely
Cambridgeshire
CB7 4EA
United Kingdom

www.itgovernance.co.uk

First published in the United Kingdom in 2013
by IT Governance Publishing.
ISBN 978-1-84928-471-4

FOREWORD

The International Standard known as "ISO9000" has been adopted by businesses and other organizations in over 170 countries around the world. There has been much conjecture and debate about this document, which is the biggest selling ISO standard since standards were created to ease international commerce. In nearly 25 years of use by various organizations, much has been said and written about the three documents collectively known as ISO9000 (ISO9000, ISO9001 and ISO9004), what they mean, and how to interpret and implement the requirements. ISO9001, the Standard which defines the requirements for an organization's quality management system and provides the criteria for certification, is the primary focus of those tasked with implementation and auditing, while the other two, ISO9000 and ISO9004 are frequently overlooked.

Although a great deal of the contents of ISO9001 would appear to those new to the subject of quality management to be "common sense," a significant number of myths have been promulgated, often from confusing what "ISO" requires, frequently based on an earlier version. These myths may have misled those who are responsible for leading implementation and may have also delayed management's understanding of the use of ISO9001 as a tool for control and improvement of business processes.

This book, then, is an attempt to expose a number of these myths and, hence, to enable a better understanding of ISO9001 by those who seek to implement or improve the quality management system of their organization.

PREFACE

The International Standard known generically as "ISO9000" has had – and continues to have – a significant effect on organizations around the world. After more than 25 years of experience by personnel in user organizations, auditors, consultants, and trainers, there are still fundamental misunderstandings of what is involved in applying the requirements contained in ISO9001.

Initially used on a "voluntary" basis for agreement on quality assurance between customers and their suppliers, the use of so-called "third-party certification," which independently audited and confirmed compliance, became a popular option for both parties. Customers saw a cost-effective alternative to maintaining supplier quality departments, instead relying on the certification bodies to maintain compliance of their suppliers' quality management systems.

Major purchasing organizations often require(d) "ISO certification" as a prerequisite for those wanting to do business as a supplier; however, they rarely provided their suppliers with practical assistance in implementation. Further, the ISO9000 guidance documents were never designed to be a road map for implementation, focusing more on improvement than on the basics.

Early adopters – particularly in the USA – coined the phrase "say what you do, do what you say, and prove it" as a description of what ISO9001 implementation involved. This may have been appropriate to the task of passing a certification audit at that point in time, but is a long way short of implementing an effective quality management system under today's ISO9001 requirements.

ABOUT THE AUTHOR

Andrew (Andy) W. Nichols has over 25 years of experience of management systems. As the Quality Manager of a UK-based NATO contractor, he was responsible for developing a Quality Assurance Program to meet the AQAP 1 Quality Assurance requirements, when supplying communications equipment (hardware and software) to the SHAPE organization. Subsequently, as Design/Supplier Assurance Manager for a well-known British-based metrology equipment company, he was on the leadership team responsible for achieving ISO9001:1987 Certification – a "first in class."

In 1990, Andy joined the first UK accredited certification body (LRQA) as a Lead Assessor, quickly becoming a supervisor and trainer, responsible for mentoring new certification auditors, leading internal sales training and also conducting ISO9000 training for clients. During this time, he performed one of the first ISO9000 Certification audits in the USA. As a consequence of that experience, he joined a small team responsible for opening the LRQA North American offices in 1992.

Shortly after arriving in the USA, he pursued a career as a consultant and trainer, with Excel Partnership Inc., a leading provider of ISO-based management systems implementation support. In this role he has delivered hundreds of ISO9000-related training courses, covering implementation, documentation, and auditing, making the material relevant to audiences ranging from shop floor personnel to CEOs of Fortune 500 Companies.

About the Author

As an instructional designer, Andy led, and contributed to, the development of "best in class" training courses for ISO9001, ISO14001, ISO/IEC 17024, QS-9000, ISO/TS 16949 and ISO/IEC 17025.

His clients have included Tellabs, Chrysler, GKN, General Motors, Visteon, Hyundai Motor Manufacturing of America, Hewlett-Packard, Dresser Industries, ANSI, the USDA, and many branches of the US Department of Defense.

During his career, Andy has held Certifications with the International Register of Certificated Auditors (IRCA) and the RABQSA organization as a Lead Auditor of Quality Management Systems. He has been a corporate member of the UK's Chartered Institute of Quality and a member of the American Society for Quality (ASQ) since 1986.

Andy joined the global certification body NQA, USA as their East Coast Sales Manager in 2008. Bringing his 25 plus years' experience to the role, he has contributed to their overall sales success.

A regular technical contributor to the largest Internet forum for Management Systems, known as the "Elsmar Cove" (*www.elsmar.com/forums*), where he is also a forum moderator, Andy can be found on the social media networking site LinkedIn, (*www.linkedin.com/pub/ andy-nichols/6/728/173*), where he is an active member of many groups and owner of the "Quality Management Discussions" group.

Andy currently lives in Michigan with his two sons and Winston the Golden Retriever.

ACKNOWLEDGMENTS

I extend my sincere appreciation to the following people:

C. Allen Powell – mentor and architect of the 5 Phase Implementation model.

David Middleton, for giving me the opportunity to pursue a career as a trainer and consultant with Excel Partnership Inc.

Jeff Monk, who trained me as a lead auditor. Jeff wrote one of the very first recognized lead auditor courses.

The Deming Organization for permission to reproduce Dr Deming's 14 Points.

John Owen, IAF Secretariat, for permission to reproduce various IAF documents.

Tim Dovan, Senior Manager, Customer Service, ANSI, for permission to reproduce various ISO documents or parts thereof.

Lt Cols Dan Mathews (USAF) and Don Lawrence (US Army) for their assistance in helping me grasp basic quality assurance practices.

The many professionals I've met during the course of my career, for providing me with the opportunities to learn from their experiences. I would also like to thank Chris Roberts-Whitehurst, Senior QMS Assessor, LRQA Ltd and Marc Taillefer, IT Service Management technical expert, consultant, trainer and coach, for their helpful contributions during the review process.

CONTENTS

Contents

INTRODUCTION

In 1987, a small, quiet and almost unnoticed event caused a revolution to occur. This revolution initially affected manufacturing organizations, mainly in the UK. However, since then it has spread throughout organizations (both profit and nonprofit) of all types and sizes around the world. The event was the publication, by the International Organization for Standardization, of the International Standard "ISO9000."

This International Standard, collectively known as ISO9000, now comprises the following documents:

- Vocabulary (ISO9000)
- Quality Management Systems Requirements (ISO9001)
- Guidance on Improvement (ISO9004).

These three are to be used by an organization wishing to adopt a recognized model on which to base its quality management system.

ISO9001 is the biggest-selling document of its type in the world. Thousands of organizations, in hundreds of countries, have implemented ISO9001. Countless auditors have been trained to audit to it, and countless documents have been written in the name of complying with it. The intention of publishing these ISO documents was to provide, in an increasingly global supply chain, a set of quality assurance requirements that suppliers could implement and, thereby, give confidence that their products and services would meet customers' needs.

Prior to the publication of ISO9000, many organizations were using quality assurance requirements in their supplier relationships. These documents were created by major purchasing organizations in the defense and automotive industries, for example. In the drug and pharmaceutical industries, regulators introduced their versions too. Suppliers to (Western) defense departments were required to comply with NATO AQAP1 or AQAP4; in the USA this was known as MIL-Q-9858A and in the UK as DEF-STAN-05/21, etc.

It is in these last documents that the first version of ISO9000 has its roots. Originally based on the British standards BS5750, parts 1, 2, and 3, which themselves were heavily modeled on the British Ministry of Defence's procurement requirements entitled DEF-STAN-05/21 (for design- and manufacturing-capable contractors) and 05/24 (for manufacturing-only contractors), ISO9000 was published in 1987 as ISO9001 (for design- and manufacturing-capable suppliers), ISO9002 (for manufacturing-only suppliers), ISO9003 (for suppliers who performed product inspection only), and ISO9004 (guidance to the other three).

In the automotive supplier base, General Motors (North America) had a supplier quality document called "Targets for Excellence," Chrysler issued the "Chrysler Supplier Quality Assurance Manual," and Ford Motor Company developed the "Q-101" specification. These "Big 3" supplier quality documents were also derived from the need for the auto makers themselves to supply their government's defense departments, and to some extent also echoed those defense contractor standards.

Major procurement organizations in other industries followed suit, including Caterpillar, John Deere, and Black and Decker, for example, passing down their requirements (often a "pass through" of their customers' quality requirements) to their suppliers.

Regulated industries also had their versions of quality assurance requirements. These were drawn up specifically for makers of medical devices and pharmaceutical drug manufacturers, particularly those in the USA with the need to comply to the Federal Drug Administration's Title 21 Code of Federal Regulation 860 (21CFR860), for example.

Pity the manufacturing organizations who supplied multiple customers in these markets! It was not unknown for organizations to create and maintain up to 14 separate quality programs to satisfy each of these customer requirements, undergoing multiple audits by supplier quality personnel, too.

Since 1987, the adoption of ISO9001 has also led to a number of industry-specific versions being developed, including:

- QS-9000 – for North American auto suppliers, which subsequently became ISO/TS 16949;
- AS9100/9110/9120 – for the aerospace industry;
- TL-9000 – for the telecommunications industry; and
- ISO/TS 29000 – for the petrochemical and gas industries.

These requirements take the basic ISO9001 requirements and add industry-specific tools and techniques to them.

At the time of writing this book, a number of ISO standards for management systems have been created to address further industry needs, including:

- ISO13485 – for the medical device industry;
- ISO27001 – for information security;
- ISO22000 – for the food industry; and
- ISO20000 – for the IT services industry.

This book represents the collective experience of implementation of quality management systems over the course of more than 25 years. Throughout there are "myth alerts" where you will find typical misconceptions relating to the requirements of ISO9001:2008, including quality management systems implementation, audits, and third-party certification.

CHAPTER 1: QUALITY – THE BASICS

Quality – just what is it?

Before we begin the dissection of the intent and content of ISO9001, it's appropriate to understand some basics which underpin the reason for the International Standard being in existence. Since ISO9000 has in its title the word "quality," a good definition of "quality" is important to understand the context of a quality management system. It is a concept which can, for some, be slippery to grasp.

There are all manner of definitions or descriptions of "quality" – some conceptual, some esoteric, and some reasonably accurate. Definitions include: "Doing it right first time," or "Selling products which don't come back, to customers who do." However, for the purposes of understanding the International Standard, we should look to the definition which is found in the "normative reference" for ISO9001: that is the Vocabulary document known as ISO9000:

degree to which a set of inherent characteristics fulfils requirements.[1]

This definition is a classical one, but may be less than helpful to those who are new to the concept. Therefore, for the purposes of understanding this book, and the application of ISO9001 to an organization, "quality" is considered as

Doing what your customer tells you they want.

---

[1] This excerpt is taken from ISO 9000:2005, with the permission of ANSI on behalf of ISO. © ISO 2013 – All rights reserved.

No surprises, no changes, no late or early deliveries – what they want, delivered time and time again. Simple, really!

Furthermore, it's going to be very necessary that the organization's management have a clear and consistent understanding of this definition, because, without it, the rest of the implementation of the ISO9001 requirements will be fundamentally flawed. The reasons will become clear later. Furthermore, it is also important that key concepts which hinge on the use of the word "quality" are also fully understood as distinct and separate – as well as being somewhat interrelated.

Those key concepts are:

- quality control,
- quality assurance,
- quality management, and
- total quality.

Quality control: or "you can't inspect quality into products!"

How "quality" is achieved by an organization has been gradually changing since World War II, particularly with the trend to mass production techniques. It was traditional, in Western manufacturing organizations, for quality control personnel – sometimes dressed in white lab coats – to inspect, check, and test products to see if they met the specification. Often, the most talented people from the manufacturing line were recruited to the Quality Control (QC) department, since they knew nearly everything about the manufacturing process – they also knew the problems which were manifest in the products. They were, simply, the best at finding the faults. This approach was costly and time consuming, and frustrated the people on the manufacturing line and their supervision. In some situations, the manufacturing departments would play games with the QC people, in order to meet production schedules and delivery dates, when products had been suspected of being rejects.

Worst of all, the QC effort directed at the product wasn't very effective in preventing defects getting to customers. A common assertion is that even 100% inspection (by people) is only 80% effective, at best.

To demonstrate this ineffectiveness, a simple test of the power of observation is used, where the reader has to count the number of times the letter "f" appears in the following sentence:

Finished files are the result of years of scientific study combined with the experience of years ...

(Answer: six)

Those who were alive during this era can often tell stories of their (or families') experiences of the defects which afflicted consumer products!

Quality control is still an aspect of ISO9001, albeit not performed the way described above. All processes need control(s) which may include measurement of process parameters and/or product characteristics and prove conformity to a specification or "quality."

Quality assurance: because assumptions are a poor choice!

As previously described in the Introduction, major purchasing organizations (mainly government entities such as departments of defense or food and drugs agencies) required suppliers to implement "quality assurance" (QA) programs. The requirements for these QA programs were published in contractually binding documents which became the predecessors of ISO9001.

The QA program requirements emphasized that suppliers had to operate the basics for assuring the quality of the products they manufactured. These basics included having the supplier create and maintain:

- a quality assurance manual,
- documented procedures/work instructions,
- a list of approved suppliers,
- measuring equipment calibration,
- controls for non-conforming product,
- training for personnel,
- a quality manager/representative,
- inspection records,
- etc.

As the use of supplier QA requirements (as they were known) became widespread, the top-level suppliers (tier 1, or primes) also started to pass them down through their supply chains. As a result, these lower-level suppliers were frequently required to implement and maintain a variety of QA programs – to meet each of their major customers' requirements. Frustratingly, although there was substantial similarity across these customer QA requirements, the supplier organizations had to maintain separate programs and documentation! Indeed, as mentioned previously, one supplier in the UK familiar to the author created and maintained 14 distinct QA manuals to satisfy this demand – despite the fact that the product was a proprietary design!

Quality management: quality doesn't just happen!

In the closing years of the twentieth century, there's been a gradual move toward the recognition that the achievement of "quality" and customer satisfaction is one of the core purposes of any organization – whether that organization is for-profit or not. With this recognition has come the understanding that management and control of the organization's processes is what delivers that "quality," from the process of developing and offering a new product into a market, to taking an order from a customer, to the delivery of that product, to any post-sales services which may also be offered. What was required of a supplier organization as QA has been enhanced and expanded to include not only the processes and activities which are closely associated with making a product, but also those which support and enable those product-related processes. This approach to quality is known today as "quality management."

We can consider the relative scopes of the three "quality" practices in the following manner:

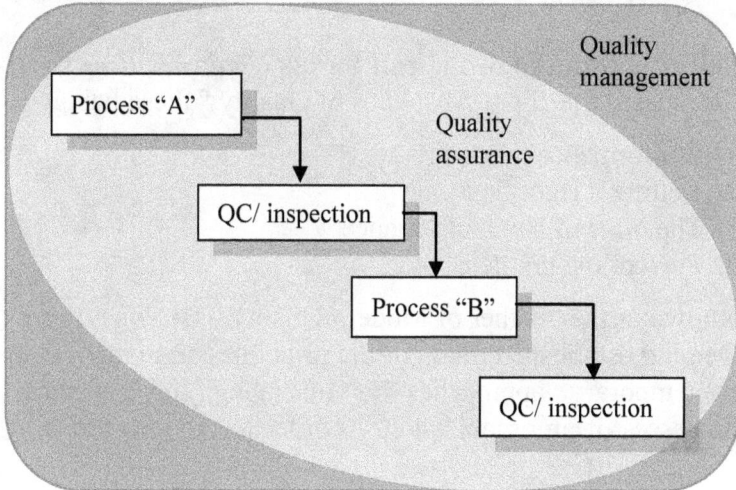

We can see that QC activities have a very specific application, often to inspect the output from a process, before the next process is carried out. QA often focuses on the elements which can also go to make the QC work more effective, for example calibration of equipment and inspection instructions; Quality Management, in turn, adds a systematic linking of those functions and other processes which go to ensure the overall effectiveness of QA and QC activities already in place.

The "total quality" movement

Between 1950 and 1980, a number of quality "gurus" came to prominence. Their work was based on the principle that inspecting quality into product isn't effective. These people shaped much of today's approaches to "quality" and to what has become known as "total quality management":

- Phillip Crosby
- Dr. W. Edwards Deming
- Dr. Armand Feigenbaum
- Dr. Walter Shewhart

Phillip Crosby, known in part for his *Quality Is Free* book, formulated the "Four Absolutes of Quality," these being:

1. Conformance to requirements
2. Getting it right first time
3. The price of nonconformance
4. Prevention, not detection

Known as the "Father of Modern Quality," Dr. W. Edwards Deming established many of the principles and practices of contemporary approaches to managing the (business) processes of an organization, in order to produce a quality product. Deming is widely credited with leading the Japanese manufacturers in rebuilding their capabilities in the years after World War II. The adoption of Deming's Management Principles by notables such as Toyota, Honda, Sony, etc. has been credited as the reason why they dominate their markets with high-quality products.

Often thought of as being the "grandaddy" of the "total quality" movement, Dr. Armand Feigenbaum conceived the idea of "Total Quality Control," which initially preceded and subsequently became what we now know as "TQM." In addition to identifying the costs associated with getting quality wrong (as did Crosby), Feigenbaum also highlighted the so-called "hidden factory" – the name he gave to the extra work done in correcting quality problems.

Dr. Walter Shewhart's focus was on the use of statistics to control manufacturing processes. His understanding of process variations and the ability to use data was inspirational to Dr. W. Edwards Deming, who used this to formulate the widely used "Plan, Do, Check, Act" cycle, also known as the "Shewhart cycle." It is this "PDCA" cycle which forms the basis of the implementation diagram developed for ISO9001:2000:

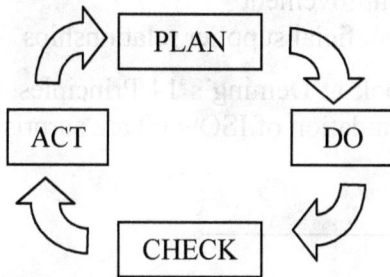

It is often suggested that none of these thought leaders in total quality management would have supported the implementation of the original ISO9000 requirements as a means to assure a quality product. The current version may, however, not be so anathema to them, since in the years leading up to the release of the year 2000 version, members of the ISO Technical Committee did much to consider and formulate how TQM concepts might be incorporated. Moving from quality assurance toward quality management made it necessary to adopt a more encompassing description of how quality and customer satisfaction is systematically and effectively achieved.

For example, the Standard now encourages organizations to consider eight quality management principles[2] when implementing their quality management system, in accordance with ISO9001.

These principles are:

1. Customer focus
2. Leadership
3. Involvement of people
4. Process approach
5. Systems approach to management
6. Factual decision making
7. Continual improvement
8. Mutually beneficial supplier relationships

If we take a look at Deming's 14 Principles, we'll see that those at the foundation of ISO9000 are surprisingly similar:

---

[2] This excerpt is taken from ISO 9000:2005, with the permission of ANSI on behalf of ISO. © ISO 2013 – All rights reserved.

1. Create constancy of purpose toward improvement of product and service, with the aim to become competitive, stay in business, and provide jobs.
2. Adopt the new philosophy. We are in a new economic age. Western management must awaken to the challenge, learn their responsibilities, and take on leadership for change.
3. Cease dependence on inspection to achieve quality. Eliminate the need for massive inspection by building quality into the product in the first place.
4. End the practice of awarding business on the basis of a price tag. Instead, minimize total cost. Move toward a single supplier for any one item, on a long-term relationship of loyalty and trust.
5. Improve constantly and forever the system of production and service, to improve quality and productivity, and thus constantly decrease costs.
6. Institute training on the job.
7. Institute leadership. The aim of supervision should be to help people, machines, and gadgets do a better job. Supervision of management is in need of overhaul, as well as supervision of production workers.
8. Drive out fear, so that everyone may work effectively for the company.
9. Break down barriers between departments. People in research, design, sales, and production must work as a team, in order to foresee problems of production and usage that may be encountered with the product or service.
10. Eliminate slogans, exhortations, and targets for the work force asking for zero defects and new levels of productivity. Such exhortations only create adversarial relationships, as the bulk of the causes of low quality and

low productivity belong to the system and thus lie beyond the power of the workforce.

11. Eliminate work standards (quotas) on the factory floor. Substitute with leadership.

12. Eliminate management by objective. Eliminate management by numbers and numerical goals. Substitute instead leadership.

    a. Remove barriers that rob the hourly worker of his right to pride of workmanship. The responsibility of supervisors must be changed from sheer numbers to quality.

    b. Remove barriers that rob people in management and in engineering of their right to pride of workmanship. This means, inter alia, abolishment of annual or merit rating and of management by objectives.

13. Institute a vigorous program of education and self-improvement.

14. Put everybody in the company to work to accomplish the transformation. The transformation is everybody's job[3].

Clearly, not all of Deming's principles were adopted! ISO9001 still maintains objectives (12) for the quality management system, but many of the eight quality management principles mentioned in ISO9001 can be seen to be aligned with Deming's.

---

[3] Deming, W. Edwards, Out of the Crisis, pp. 23-24, © 2000 Massachusetts Institute of Technology, by permission of The MIT Press.

ISO9000: a legend in its own lifetime?

Today, after nearly 25 years of implementation and certification experiences, all over the world, many myths and legends have grown and been promulgated surrounding the use of ISO9000, including:

Say what you do, do what you say

and

ISO Certification allows you to make concrete life jackets (a similar quote was printed in USA Today in 1998).

Many of these myths have delayed management's interest in utilizing ISO9000 as a tool to improve the way business is conducted, to drive the result that customers are happy, which in turn leads to increased sales, so that management can focus on reduced waste and higher efficiencies through implementing an improved management system of processes and controls.

Today, various Internet groups and forums are filled with questions and comments which have been posted by users concerning all manner of ISO9000-related issues. Browsing these posts shows that many of the original myths and legends which began to pervade implementation of ISO9000 requirements back in the 1990s are still alive and kicking today.

Perhaps a contributory factor in attracting so much misunderstanding of the purpose behind the International Standard is that there was no model or other description of what a quality management system should be when implemented.

Early versions of the ISO9001 requirements heavily emphasized documented procedures for each of the 20 "elements" or clauses – which could easily be aligned to individual functions or departments – with the result that implementing organizations often built a quality management system of paperwork for each function. In doing so, they often failed to address the quality problems resulting from interdepartmental "dysfunction." This "silo approach" was, in part, the reason for the "Say what you do, do what you say" myth.

In the year 2000 this was corrected when the following diagram was incorporated to ISO9001[4]:

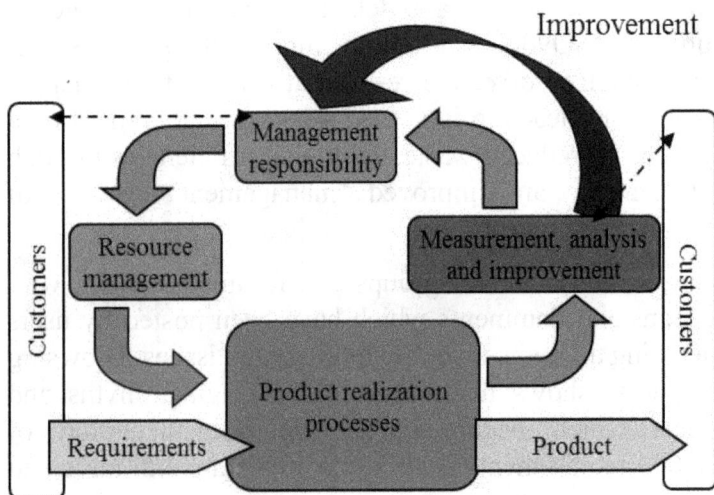

The diagram is intended to be read thus:

Management's responsibility is to identify the quality management processes, set objectives, and define quality

---

[4] Adapted from ISO 9001:2008, Figure 1, page vi, with the permission of ANSI on behalf of ISO. © ISO 2013 – All rights reserved.

policy, based on understanding their marketplace and their customers' needs and expectations.

Resources are identified and provided to deploy to work on the processes of the quality management system, working to achieve the stated objectives and policy.

The "product realization" part of the diagram represents those processes which accept customer requirements and transform them into a deliverable product to the customer.

"Measurement, analysis, and improvement" takes a look at both product conformity and process performance (compared to the set objectives) – including the all-important feedback from customers – through analysis of the associated data, to identify opportunities to improve product and/or the processes of the quality management system.

This information is fed back to management, so that they can review it and make decisions on whether customers were satisfied, on whether products conformed, and on whether processes were being followed and were effective in meeting objectives. Any improvements can be supported with the necessary resources ... and so the cycle starts again.

In July 2009, the International Organization for Standardization (ISO) and the International Accreditation Forum (IAF) released a joint communiqué on the expected outcomes of ISO9001 and accredited certification. It states:

Expected Outcomes for Accredited Certification to ISO9001 (from the perspective of the organization's customers)

"For the defined certification scope, an organization with a certified quality management system consistently provides products that meet customer and applicable statutory and regulatory requirements, and aims to enhance customer satisfaction."

Notes:

a. "Products" also include "services."

b. Customer requirements for the product may either be stated (for example in a contract or an agreed specification) or generally implied (for example in the organization's promotional material, or by common practice for that economic/industry sector).

c. Requirements for the product may include requirements for delivery and post-delivery activities.

What accredited certification to ISO9001 means

To achieve conforming products, the accredited certification process is expected to provide confidence that the organization has a quality management system that conforms to the applicable requirements of ISO9001. In particular, it is to be expected that the organization:

A. has established a quality management system that is suitable for its products and processes, and appropriate for its certification scope

B. analyzes and understands customer needs and expectations, as well as the relevant statutory and regulatory requirements related to its products

C. ensures that product characteristics have been specified in order to meet customer and statutory/regulatory requirements

D. has determined and is managing the processes needed to achieve the expected outcomes (conforming products and enhanced customer satisfaction)

E. has ensured the availability of resources necessary to support the operation and monitoring of these processes

F. monitors and controls the defined product characteristics

G. aims to prevent nonconformities, and has systematic improvement processes in place to:

1. Correct any nonconformities that do occur (including product nonconformities that are detected after delivery)

2. Analyze the cause of nonconformities and take corrective action to avoid their recurrence

3. Address customer complaints

H. has implemented an effective internal audit and management review process

I. is monitoring, measuring and continually improving the effectiveness of its quality management system

What accredited certification to ISO9001 does mean

1) It is important to recognize that ISO9001 defines the requirements for an organization's quality management system, not for its products. Accredited certification to ISO9001 should provide confidence in the organization's ability to "consistently provide product that meets customer and applicable statutory and regulatory requirements." It does not necessarily ensure that the organization will always achieve 100% product conformity, though this should of course be a permanent goal.

2) ISO9001 accredited certification does not imply that the organization is providing a superior product, or that the product itself is certified as meeting the requirements of an ISO (or any other) standard or specification.

It can be seen that the myth of the concrete life jackets is just that – a myth! Since in several of the preceding paragraphs there are many references to meeting statutory and regulatory requirements, as well as determining customers' needs and expectations, a concrete life jacket couldn't be designed, tested, and produced, or delivered to a customer. There are numerous regulations which require a life jacket to float – with an adult strapped into it – and product testing would have verified the ability of the jacket to float (in accordance with the regulations). These requirements would have been factored in to a number of key requirements of ISO9001 and would be necessary components of the life jacket manufacturer's quality management system!

The following chapter discusses other myths of individual ISO9001 requirements.

CHAPTER 2: ISO9001 AND ITS MYTHS

In this chapter, we'll take a look at the various sections and requirements of ISO9001, in depth, including some of the more common myths associated with each requirement encountered. It is recommended that a copy of the ISO9000 standards are obtained since the full requirements are not contained in these texts.

In addition, practical guidance is given – often through concrete or everyday examples – of what each requirement means, based on several years' hands-on experience of designing, documenting, implementing, and auditing a few hundred management systems.

ISO9001 has an Introduction and four key sections:

1. Scope
2. Normative references
3. Terms and definitions
4. Quality management system

Rarely are all the requirements of this 27-page document ever fully read and completely understood. Instead, most readers tend to focus on the Section 4 requirements, mainly because that's where the auditable (for certification) requirements are detailed. Sadly, that's like sitting down to a four-course meal and going straight for the coffee and mints! The soup, salad, entrée, and dessert have all been missed and the fullest benefit of the dining experience somewhat lost – how bizarre!

It was only when the ISO9001:2000 version was published that any thought was really given to a description of how all the requirements of the International Standard were supposed to come together as the basis of the organization's quality management system. Despite volumes of guidance and a significant number of books being published on the subject of implementing and auditing ISO9001!

Introduction

The introduction to ISO9001 talks about the adoption of a quality management system as being a strategic decision. It then goes on to list the influences on the design of the quality management system, including:

- the organization's environment, and any changes and risks associated with that environment, and
- the organization's varying needs, particular objectives, products, processes, size, and organizational structure.

It is interesting to note that the word "risk" is introduced at this point – and almost never repeated again (see also the chapter on the future of ISO9001 below).

Scope (Section 1)

Although Third Party Certification and ISO9001 are often discussed synonymously, the scope statement doesn't clearly indicate that relationship. Instead it leaves it to the implementing organization to decide how it will demonstrate compliance. Indeed, it goes further than the basic original premise of ISO9001, that of simply meeting customer requirements, to making improvements to the Quality Management System, thus:

The requirements of ISO9001 are for use by an organization wanting to:

    a) demonstrate its ability to consistently provide product that meets customer and applicable statutory and regulatory requirements, and

    b) aims to enhance customer satisfaction through the effective application of the system, including processes for continual improvement of the system and the assurance of conformity to customer and applicable statutory and regulatory requirements.

Two notes follow defining the term "product" as only applying to

    a) product intended for, or required by, the customer

    b) any intended output resulting from the product realization process

Of course, while Certification by a Third Party isn't explicitly mentioned, this is the most common method chosen to meet a) above, often at the command/request of customers.

Normative references (Section 2)

Since the ISO9001 Standard is written to have the same meaning in many languages, it's important that words which represent key concepts on which a quality management system are built have common definitions. Without a clear understanding of these key words, the design, implementation, and improvement of a quality management system is not going to be as effective.

A quick look at discussions on various Internet forums will quickly reveal how important it is to have common understanding of terminology. One example, which we will explore in a later section, is that of defining "competence." A quick look at a variety of (online) dictionaries shows quite a number of definitions which may not fully address the intent, in the context of ISO9001. This key definition isn't found, however, in the actual ISO9001 standard itself, but in ISO9000, which is the so-called "normative reference," containing the terms and definitions necessary to guide the reader through the ISO9001 requirements.

Competence, which is the subject of clause 6.2.2 of ISO9001, is defined in the vocabulary document[5] – ISO9000 – as "The demonstrated ability to apply skills and knowledge." In the context of the requirements specified, it is easier to see that once competences have been defined and demonstrated, some actions may be necessary, including training, to address any shortfalls identified. Such a process will, now a better understanding has been gained, lead to time and money being saved, compared to the conventional training programs which might typically have been developed.

Terms and definitions (Section 3)

Since its release, ISO9001 has employed the terminology of "product" as the resulting output of the quality management system. The background to the development of ISO9001 was, after all, the military hardware-oriented defense

---

[5] This excerpt is taken from ISO 9000:2005, with the permission of ANSI on behalf of ISO. © ISO 2013 – All rights reserved.

standards. From the feedback of users in the years since 1987, it became obvious that organizations which provided services (often alongside their product) had struggled to implement the requirements of ISO9001, although having a formal quality management system was found to be useful and was often a customer requirement. The need for fewer product-oriented terms was called for.

Under this heading, a clear statement is made that where the word "product" is mentioned, users may substitute "service" if this is applicable to their needs. A reference is also made here to the use of ISO9000 for the other definitions.

The quality management system (Section 4)

Myth alert!

"We'll buy an 'ISO system' from the Internet! Have a consultant write it for us."

Section 4 defines the overall purpose of the quality management system, and describes it in short form, which can be laid over the "Plan, Do, Check, Act" (PDCA) cycle.

Beginning with the statement that the organization shall establish, document, implement, and maintain a quality management system and continually improve its effectiveness, in accordance with the requirements of the International Standard, we can see the beginnings of the PDCA cycle.

This outline is further expanded on in the following texts:

- Plan:

o Determine the processes needed for the QMS and their application throughout the organization.

o Determine the sequence and interaction of these processes.

o Determine the criteria and methods needed to ensure effective operation and control of the processes.

- Do:

o Ensure the availability of resources and information necessary to support <u>operation</u> and monitoring of the processes.

- Check:

o Monitor, measure as applicable, and analyze the processes.

- Act:

o Implement actions to achieve planned results and continual improvement of the processes.

Toward the end of the text, the practice of outsourcing processes is mentioned. Three notes below the text provide guidance on what an "outsourced" process is – something needed for the quality management system but performed by an external party. In simple terms, control is required to be exercised over the external source. Controls can be varied and range from source inspections/QC at the supplier to incoming QC at the organization's premises.

Many (manufacturing) organizations will have in place some provision for supplier quality controls as part of a "receiving" function. Care must be taken, as with all such controls, that they are effective in reducing the risks of defects entering. See also the clauses dealing with Purchasing – 7.4.1 and 7.4.3. Care must be taken when outsourced processes include those which cannot have their quality subsequently verified by inspection or testing.

Myth busted!

An effective quality management system cannot be created by anyone else, without the active participation of the organization. That you can buy a compliant "off-the-shelf" quality management system is, in fact, a fallacy. What may be offered for purchase is simply a set of documentation, which may need more work to modify than if the same were created from scratch. A documented, effective management system takes work on the part of the organization.

Myth busted!

Hiring a consultant to write documentation achieves the same results as above and is often costlier. A consultant can be used to guide an organization in creating a documented system and provide options and tools to meet the ISO9001 requirements, but cannot effectively document a quality management system – which has ownership by the organization's personnel. Once the consultant walks away, any issues will be "their fault." An effective consultant knows how to guide and facilitate the organization's personnel in creating a quality management system which they are prepared to own and, therefore, to improve.

Documenting the quality management system (subsection 4.2)

Myth alert!

The QMS is documented in 4 "levels" of documents.

All procedures need work instructions.

A process should have a documented procedure to describe it.

All ISO9001 requirements should be addressed by documented procedures.

Only six documented procedures are required.

In the 1987 and 1994 editions of ISO9001, there were multiple requirements where a documented procedure was specified and in the process control section (subsection 4.9) a reference was made to the use of work instructions needed "where the absence would adversely affect quality." This was in part due to the roots of the Standard being in the military contractor requirements documents. Possibly, from this implied convention, organizations took a tiered approach to documenting their quality management system, like this:

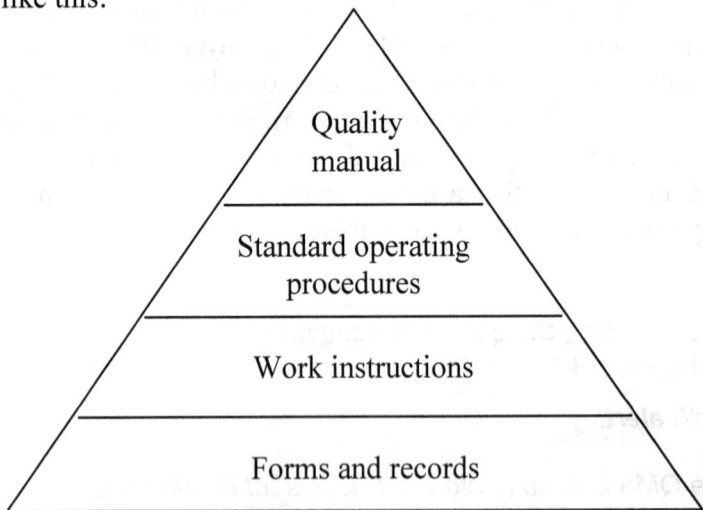

Quality manual

Standard operating procedures

Work instructions

Forms and records

Although this visual metaphor served some purpose, the major changes which were made to the requirements of ISO9001, in the year 2000, reduced the need for documented procedures. Instead, the Standard requires organizations to define their processes and associated controls, which in turn frees them to design their documentation in complete freedom from paradigms like the document pyramid.

In short, organizations must have a "documented system," not a "system of documents."

General Requirements (clause 4.2.1)

As has been discussed, no model is given for how the documentation should be assembled to form the quality management system. This requirement merely outlines what is required and what may be included:

- "documented statements of quality policy and quality objectives"
- "a quality manual"
- "documented procedures and records required by the Standard"
- "Document and records determined by the organization to be necessary to ensure the effective planning, operation and control of its processes."

This final sentence gives a great deal of flexibility for the organization to determine what's necessary to be documented. Some guidance notes[6] follow these texts

---

[6] Adapted from ISO 9001:2008, Note 3, page 3, with the permission of ANSI on behalf of ISO. © ISO 2013 – All rights reserved.

which indicate the considerations which might be made before creating further documents. The notes include the influences on the extent of documentation, which is dependent upon:

- the size of the organization and the type of its activities,
- process complexity and interactions, and
- the competence of personnel.

To illustrate this, it's relatively easy to consider a kitchen in any household. We could ask, would it be expected to see the process of preparing and cooking scrambled eggs being written in some form of document? Probably not, since it's a simple process, usually performed by competent cooks.

Now, let's consider a more complex task, like making a cake containing many ingredients – a fruit cake, for example. Unless the baker is preparing and baking this cake on a regular basis, the ingredients and method may need to be written down for reference. The basic cake mix – six measures of sugar, six of butter, and six of flour, and three eggs – may well be remembered, but the other ingredients and proportions may not be recalled as accurately. Cherries? Raisins? Nutmeg? Almonds? How much to add and when?

Furthermore, while combining the eggs into the beaten butter and sugar mixture, it is common for the mixture to curdle, which can affect the resulting cake. It's a situation which can be recovered by adding some flour to the mix. Rarely does a recipe ever mention this technique, yet a competent cook will have it memorized and know exactly when to employ it!

A balance can be struck between capturing those key process controls and the basics which make a person

competent. In addition, as with recipes which are handed
down through families, documenting processes, controls,
and activities also helps preserve knowledge and provides a
basis for developing competence in the next person.

The quality manual (clause 4.2.2)

A quality manual was never required in previous versions
of ISO9001, yet, despite this fact, most organizations
created them as a means to comply with the Standard. In
the 2008 version, the content of a quality manual is
described as being:

- the scope of the quality management system, plus any
  exclusions and the justifications for them;
- the documented procedures – or reference to them; and
- a description of the sequence and interaction of the
  processes of the QMS.

Myth alert!

*Quality Manuals must address all the clauses of the
Standard.
The more pages, the better!*

*Using the diagram from ISO9004 can serve as the
"sequence and interaction of the processes of the QMS."*

As with the other documentation myths, the quality manual
is fabled for the need to address every clause or
requirement of the Standard. Therefore, quality manuals
end up being many pages long (often 25–40) and, in many
cases, thinly disguised reproductions of the exact text, cut
and pasted right from the ISO9001 Standard. As a
consequence, the terminology of the ISO document is used,

which is alien to most organizations, especially since "product realization" isn't the normal description of "design engineering," or "manufacturing," "machine shop," "service center," etc.

A quality manual is often created along these lines, to be used exclusively by an external auditor (certification body or supplier)! Since the language found in the ISO standard is somewhat "foreign" and the structure doesn't reflect the actual business processes of the organization, it's no wonder that the resulting document is rarely used by anyone else!

If someone were inquisitive enough to take a peek inside, they'd probably fall asleep from boredom after a few pages, so maybe they should have a warning about the soporific effect on the front cover:

"Do not operate heavy equipment while reading"

The requirements of ISO9001 which describe the contents of a quality manual, plus a few others which would be appropriate content, can be very effectively presented to any reader in a tri-fold (if a paper copy is chosen) somewhat like a marketing brochure for a product!

In actuality, it's not the number of pages which is the measure of the success of a quality manual. It's its acceptance by and value to users. A useful quality manual should be a "road map" for how quality is managed within the organization. It might be all that's required to make the quality manual helpful, is a definition of the quality policy, related objectives, the processes and procedures that the organization has set out to achieve those objectives, and some other details, perhaps including an organization diagram and an indication of who in the organization holds the responsibility of the "management representative," instead of something which gathers dust until an auditor requests it!

It is true to say that, for smaller organizations, the quality manual may be a compilation of both the requirements taken from ISO9001, clause 4.2.2 and also the six documented procedure requirements:

- document control,
- records control,
- control of nonconforming products,
- internal audits,
- corrective actions, and
- preventive actions.

As with other requirements of ISO9001, no one part should be read in isolation from any other parts which are related. If we review the documentation requirements in subsection 4.2, it shows that other documents are also needed, in addition to the quality manual and documented procedures, including:

- quality policy;
- quality objectives; and
- other documents needed to ensure effective planning, operation, and control of processes.

If we take these documents and consider adding them into the organization's quality manual as an appropriate place for them to exist, and since the documented procedures may only be one or two pages in length, it can be seen that a small organization's quality manual (simple in scope, such as for a component machining shop) might only need to be 20 to 25 pages (in hard copy)!

An organization may elect to utilize technology to document its quality management system, and, of course, the same constraints on pages don't necessarily apply, although the ability to navigate through a quantity of pages – even on an intranet – can still detract from a user's experience.

Myth busted!

A quality manual needs to contain only those requirements of ISO9001 which are applicable to the organization.

Myth busted!

An effective quality manual is determined by the user, not by size, content, pages, or elements of the Standard.

Myth busted!

The diagram shown in ISO9004 (Figure X) is a model of how quality management systems are to be operated, not what the organization's processes look like or how they interact. A description of the sequence and interaction of processes should be determined by the organization's management. (This is often hard work for them, since it's rarely ever been done before, so expect disagreement!)

Control of documents (clause 4.2.3)

Myth alert!

Forms don't need to be controlled.

Documents should be indexed to ISO9001 clauses.

Documents must be signed to show review and approval.

Documents must be numbered.

Documents which aren't controlled must be stamped "Uncontrolled. For Reference Only."

A procedure for procedures is required.

The control of documents, although not usually institutionalized in many organizations, doesn't have to be complex, or a nightmare. The requirements[7] are fairly simple and straightforward:

- A documented procedure is required to control documents, which includes the following:
  o Approving documents for their adequacy before use.
  o Reviewing and updating, when necessary, and reapproving them.
  o Identifying changes and current revision status.
  o Making available relevant versions at points of use.
  o Make sure they remain legible and identifiable.
  o Preventing the use of obsolete documents and ensuring they're marked if kept.
  o Ensuring that documents which are produced outside the organization are identified and that distribution is controlled.

This last requirement is frequently either overlooked or misunderstood. Simply put, there is often a need to use documents created outside the organization – examples include material specifications, customer drawings, regulations applicable to product design safety, machine

---

[7] Adapted from ISO 9001:2008, section 4.2.3, page 3, with the permission of ANSI on behalf of ISO. © ISO 2013 – All rights reserved.

maintenance manuals, and so on. The basis for using such externally produced documents (what the Standard refers to as "of external origin") is that the correct (often most recent) one is available for use. Where national or international standards are employed, an updating service can be contracted to provide details, usually annually, on any changes so that the organization may decide to purchase what they need. When implementing document controls, it is worthwhile to inventory any externally produced documents as a way to discover who has them, and to ensure that they have the correct version, get updates, etc.

Much of the necessary control of documents can best be achieved by the use of a document change control (DCC) form. If designed correctly, this vehicle can provide:

- a means to request changes to documents, through the new, revised, and obsolete "phases";
- a record of the background to the need for the above;
- a record of approval and of review comments; and
- instructions for implementation of the above.

It is difficult, undesirable, and unnecessary to incorporate the above in individual documents since it adds to the bureaucracy described earlier, often adding detail but little or no value to the user.

An example DCC form is shown in *Appendix 1*.

Myth busted!

Forms need to be controlled – they contain vital data on products and process and ultimately become records.

Myth busted!

Quality management systems documents are needed by the organization for a variety of reasons, some of which may have no direct correlation to an ISO9001 requirement. Experience shows that complex numbering systems cause a lot of work, are of little relevance to users, and can even cause extra work when standards are changed to align references when it's not really required.

Controls of records (clause 4.2.4)

Myth alert!

Records must be signed as approved.

Records should be retrievable on demand.

Records should be revision-controlled.

Only the records stated in ISO9001 are needed.

Plans and schedules are records.

Disposition of records means disposal.

Records should only be retained for three years.

Old procedures and work instructions are a record of what happened.

Records are defined in clause 4.2.3 as being a "special type of document" – which, while giving an indication that they aren't to be "lumped in" with all documents, doesn't give much of a clue as to why they are considered special. Simply put, these records demonstrate "conformance to requirements" and also "the effective operation of the Quality Management System."

Reading through ISO9001, a number of requirements indicate an associated record. Alternatively, ISO has a guidance document titled "Introduction and Support Package: Guidance on the Documentation Requirements of ISO9001:2008," wherein Annex B is a list of records required by ISO9001:2008.

This is their list of required records[8]:

- 5.6.1. Management reviews.
- 6.2.2. Education, training, skills, and experience.
- 7.1.d. Evidence that the realization processes and resulting product fulfil requirements.
- 7.2.2. Results of the review of requirements related to the product and actions arising from the review.
- 7.3.2. Design and development inputs relating to product requirements.
- 7.3. Results of design and development reviews and any necessary actions.
- 7.3.5. Results of design and development verification and any necessary actions.
- 7.3.6. Results of design and development validation and any necessary actions.
- 7.3.7. Results of the review of design and development changes and any necessary actions.
- 7.4.1. Results of supplier evaluations and any necessary actions arising from the evaluations.

---

[8] Adapted from ISO 9001:2008, Annex B, pages 20 – 25, with the permission of ANSI on behalf of ISO. © ISO 2013 – All rights reserved.

- 7.5.2. Records, as required by the organization, to demonstrate the validation of processes where the resulting output cannot be verified by subsequent monitoring or measurement.
- 7.5.3. The unique identification of the product, where traceability is a requirement.
- 7.5.4. Customer property that is lost, damaged, or otherwise found to be unsuitable for use.
- 7.6.a. Basis used for calibration or verification of measuring equipment where no international or national measurement standards exist.
- 7.6. Validity of the previous measuring results when the measuring equipment is found not to conform to requirements.
- 7.6. Results of calibration and verification of measuring equipment.
- 8.2.2. Internal audit results and follow-up actions.
- 8.2.4. Indication of the person(s) authorizing release of product.
- 8.3. Nature of the product nonconformities and any subsequent actions taken, including concessions obtained.
- 8.5.2. Results of corrective action.
- 8.5.3. Results of preventive action.

These requirements, depending on the applicability of each to the organization's quality management system, may represent the 21 (minimum) records that are to be

- identified,
- stored,
- protected,
- retrieved,
- retained, and

- dispositioned

according to the requirements of this ISO9001 section.

It is entirely possible that other records should be added to the list, particularly those which might be necessary to comply with regulatory or statutory requirements. Even if the organization is not working in a regulated environment, it is highly likely that other records may be needed to demonstrate that processes are being controlled in order to meet requirements. For example, if a work environment needs specific conditions of temperature or humidity, or if clean room conditions to ISO14644-1 are required to ensure that product quality is maintained, then records of the relevant measurements of those parameters should be identified.

Records are the basis of the data needed by the organization's management to guide their decisions related to the following aspects of the quality management system:

- the suitability of the quality policy,
- the achievement of quality objectives,
- customers' perception of requirements being met,
- conformity of products, and
- supplier performance.

Records may also provide valuable evidence protecting the organization from legal action. Original product design records have been successfully used to pursue patent infringements by a competitor organization. Product liability lawsuits can require the production of records of product design/development and manufacture/testing to demonstrate due diligence. Advice should always be sought from legal counsel regarding the retention period of records. In addition, the myth of the disposal of records should be addressed by nominating certain individuals to decide the fate of records. Important records might never be disposed of; instead, they may be saved from the trash, from erasing, burning, or shredding, by sending them to an off-site storage facility specializing in records storage.

In summary of the two requirements for control of documents and records, a simple aphorism might be:

Documents say "do," records say "done."

Management responsibility (Section 5)

Management commitment (subsection 5.1)

Myth alert!

Signing the quality policy is a demonstration of commitment to ISO9001 by the top management.

This requirement of ISO9001 states that "Top Management" must demonstrate their commitment to the development and implementation of the Quality Management System and continually improving the effectiveness[9]. Referring back to the eight quality management principles covered earlier, the role of leadership in ensuring an effectively implemented quality management system is very clear. But what is meant by commitment and how can an organization demonstrate it?

---

[9] Adapted from ISO 9001:2008, section 5.1 page 3, with the permission of ANSI on behalf of ISO. © ISO 2013 – All rights reserved.

We can often see clear examples around us of what it takes to be committed to something. Take, for example, the commitment of parents to their children's sports activities. These parents demonstrate support and involvement through the resources they supply:

- time,
- budget,
- knowledge,
- involvement, and
- encouragement.

Top management display their commitment through active involvement in resourcing the implementation of the quality management system. They devote time to participation in setting quality policy; establishing and communicating quality objectives; and communicating the importance of meeting customer, regulatory, and statutory requirements and management reviews. Top management take an interest in quality matters, in providing the resources necessary for corrective action and for improvements.

Customer Focus (subsection 5.2)
Myth alert!

Only people who deal directly with the customer, like Sales or Customer Service, need to comply with this requirement.

People on the shop floor don't affect what the customer is actually sold.

A simple ISO requirement, but often overlooked, is based on the fact that nearly everyone in an organization is there to meet, or support meeting, customer requirements. It's up to the organization's management to ensure that people – and processes – deliver.

When talking to people about their jobs – whether casually or even as part of an audit – it's always interesting to learn how well people can describe how their work affects the (ultimate) customer – or even the internal customer!

Quality Policy (subsection 5.3)

Myth alert!

The quality policy must be remembered/recited by all personnel.

The quality policy must be signed by the top manager of the organization.

Policies are needed for each requirement of ISO9001.

Often confused with an organization's "vision" and "mission" statements, the requirement of ISO9001 to define its quality policy is frequently not given as much thought and deliberation as is really necessary. Typically, this manifests itself in the levels of commitment made in examples of quality policy statements:

"It is the policy of our company to meet and exceed customers' requirements."

"Our organization's quality policy is to exceed our customers' expectations and deliver on time every time."

While these are admirable intentions, in and of themselves, however, such statements have three inherent issues.

First, for an organization to be able to determine a customer's expectations – which are commonly not stated – and to exceed them is likely to be a practical improbability.

Second, statements such as these are somewhat "lofty" and leave little opportunity for improvement.

An example quality policy might state:

"We're committed to achieving customer satisfaction, which we'll obtain by meeting their requirements for our products, produced to specification, every time. We will improve what we do through the use of our quality management system."

It can be seen that there are key words which provide a platform for establishing quality objectives:

- "customer satisfaction,"
- "meeting requirements,"
- "producing to specification, every time,"
- "improvement of what we do."

Finally, because the quality policy is supposed to be a framework for establishing quality objectives, any policy which demands such high performance will result in those objectives demanding high performance from the quality management systems processes and, therefore, the personnel who operate them. It's well known in industrial psychology that setting unattainable goals can be demotivating for people. Setting goals and objectives which are achievable, supplemented by some "stretch" goals, would be entirely appropriate in most cases.

Planning (subsection 5.4)

The requirements for planning are fairly insignificant, at first sight, but, when we view them with other related requirements, we can see the formation of clear relationships:

- The setting of Quality Objectives, flowing down from the Quality Policy, as a means to measure and monitor process performance.
- Planning as a means to ensure that the established quality management system is used as a means to control changes, thereby ensuring continuity of effectiveness of process performance.

Quality objectives (clause 5.4.1)

Having determined a suitable quality policy for the organization, top management are required to ensure that objectives are established at "relevant functions and levels within the organization." These objectives have to be measureable and consistent with the quality policy.

As we've previously covered, the quality policy should provide a framework for establishing quality objectives. Management should establish (if they haven't already) what the current measurements and performance are for these parameters. Having determined these, agreement should be sought as to a set of goals (and, possibly, "stretch" goals) relative to the subjects. For example:

- Current customer satisfaction may be at 87%. A goal may be established at 94% for the coming period, with a stretch goal of 97%.
- Meeting requirements may be focused on delivery performance, which may be at 93% on time. A goal may be set at 97%, with stretch goal of 99%.
- Producing to specification may be determined by "first pass yield" (FPY) or "first time through" (FTT) for

products. Such a measure is often used in a manufacturing environment; however, it can equally apply to each stage in the product development process, or even in the processing of responses to customers' requests for quotes.

- FPY or FTT can be measured by the number of product quality rejects which cause less than 100% results from the manufacturing process(es).

- Objectives may be set at a functional level – for example, an FPY or FTT of 97% – and may then be translated to a specific process/product level which can contribute to this.

Quality management system planning (clause 5.4.2)

Quality management system planning can be achieved in different ways, depending on the needs of the organization. In some industries it is common to have a formally defined quality planning process – the automotive industry uses a process known as "Advanced Product Quality Planning," or APQP. This process calls for a number of specific quality tools to be used, in sequence, by a cross-functional team of representatives from a number of functions.

Of course, ISO9001 allows the implementing organization to decide how quality planning is accomplished. For some, probably small(er), organizations, this requirement is met through implementing the quality management system in a set sequence:

1. review of customer orders,
2. production scheduling/planning,
3. production monitoring,
4. etc.

Management may also conduct regularized operational reviews to establish a business plan and estimate revenues and a budget for the coming financial period, setting objectives and then monitoring the progress of their achievement. These reviews would form a suitable vehicle for planning changes which can affect the quality management system, assigning responsibilities and controls needed, product and process performance monitoring, and even internal audits to ensure the changes carried out are under control and avoid negative impacts. The result of this activity may also result in a documented "quality plan."

Responsibility, authority and communication (subsection 5.5)

Responsibility and authority (clause 5.5.1)
Myth alert!

An organization chart is needed.

Quality personnel must report directly to the top management person.

Job descriptions are necessary.

Everyone needs to know who holds the position of "ISO management representative."

It's often said that "everyone's responsible for quality, so nobody is." The requirement of the ISO standard is an almost insignificant single sentence.

In general terms, job or position descriptions exist in the majority of organizations, frequently to address employment laws, make hiring or even termination easier, etc. Although such job descriptions give details of high-level responsibilities, descriptions of day-to-day work aren't always included, since this can vary.

Furthermore, the levels of authority – which can be of significance – are rarely included in job descriptions for the majority of personnel. Typically, authority is assigned for people with financial and budgetary responsibility. Assigning authority for product quality and process control is of particular importance, and can be accomplished throughout the documented quality management system, particularly in key procedures, for example:

- approval of documents,
- disposition of records,
- disposition of nonconforming products, and
- taking corrective action.

Authorities should be defined throughout the quality management system where people will need to make commitments to supply customers, maintain regulatory compliance, and approve designs/products for release and so on.

Defining authorities for key aspects of the quality management system, if overlooked, can have potentially disastrous consequences. Unqualified persons approving reworked, previously defective products' being shipped to customers can have life-threatening results!

Management representative (clause 5.5.2)
Myth alert!

The organization must appoint a "quality manager."

The management representative must be an employee.

It's the management representative's job to do "ISO."

The role of the management representative is an important one in both establishing and maintaining the quality management system in effective compliance with the requirements of the International Standard, customers' needs, and any regulations which may apply.

Often, the role of the management representative is as the architect of the quality management system. It is their role to lay down the plans for how the system is to be designed and implemented, and how the various requirements of ISO9001 (which are applicable), and any other applicable regulations, are met.

Internal communication (clause 5.5.3)

Communication process(es) are to be used to communicate the effectiveness of the quality management system. Having established quality objectives at various levels of the organization, it would make sense to keep people informed about the progress being made toward them, so that they can keep a focus on what's important. Management may elect to employ a "balanced scorecard" to communicate throughout the organization. The key areas of focus are often:

- finance,
- customers,
- internal processes, and
- learning and growth.

By reference to the requirement for quality objectives to be established (clause 5.4.1) at relevant functions and levels, it can be seen that these can be mapped to the balanced scorecard topics.

Individuals' objectives in support of the organization's overall objectives can be communicated through personal (annual) performance reviews, daily shift briefings, team notice boards in the workplace, and many other techniques. One desirable outcome of this communication should be that people involved in meeting customer requirements should be able to describe their role in the overall achievement of objectives and current performance.

Management Review (subsection 5.6)
Myth alert!

Management review has to be a meeting.

It's the management representative's job to present the review.

One review a year is all that's needed.

We have weekly/monthly quality meetings – they're our management review.

Management review of the implementation of the effectiveness of the quality management system is supposed to consider the following (in no particular order):

- status of actions from previous reviews,
- audit results,
- customer feedback,
- product conformity,

- process performance,
- corrective action status,
- preventive action status,
- changes affecting the quality management system, and
- opportunities for improvement[10].

One way to consider management reviews of the quality management system is to think of it as a navigating exercise on board a yacht! When starting out on a voyage, it's wise to start out with a plan (map) and some objectives, a way to measure progress, clear responsibilities for the crew, someone in authority (a "captain"), and resources (trained crew, food, water, etc.) to support getting to the planned destination on schedule.

Once a course has been determined and plotted, everyone can go about their various processes and tasks of sailing the yacht. To ensure that the voyage ends at the planned destination, it will be necessary for the captain to confirm that the vessel is still headed in the planned direction and that weather and tides haven't moved the yacht from the intended course. In addition, progress should be measured to ensure that the objectives of the voyage are kept in sight and are still achievable. Consideration may also be given to the available resources and whether they are also sufficient to support the objectives. Course corrections may be required and speed adjustment may be necessary to ensure the journey is completed on time.

Clearly, consulting the map, checking progress, or even waiting until a destination is reached would be ineffective

---

[10] Adapted from ISO 9001:2008, section 5.6.2, page 5, with the permission of ANSI on behalf of ISO. © ISO 2013 – All rights reserved.

in ensuring objectives are met; the correct destination may not be reached, resources may be expended before the voyage is completed, arrival may be delayed, and measurements taken during the voyage may not have been accurate.

Of course, at the end of the voyage, it would be helpful to "take stock" of the situation to see if there were any lessons to be learned. In addition, it would be very prudent to ensure that the plan is consulted at least once or twice while en route, in order to discover any deviations before they put the voyage in jeopardy.

The same applies to management's review of the quality management system. Once a year is of no value at all, twice won't tell you too much more. Quarterly would give opportunities to manage the performance of the processes, taking a timely look at the need for correction or improvements along the way. It can be seen that an effective review of the quality management system is the cornerstone to maintaining and improving its effectiveness.

Resource management (Section 6)

Provision of resources (clause 6.1)

Resources are required by ISO9001 to achieve essentially two aspects of the quality management system:

- implementation, maintenance, and improvement of the QMS; and
- enhancing customer satisfaction by meeting requirements.

The resources needed for these include the human ("software") and the infrastructure ("hardware").

Competence, awareness and training (clause 6.2.2)

Myth alert!

Everyone needs training on "ISO."

You can "grandfather" people into their jobs.

One area of implementation where a significant amount of money can be saved is in NOT training people! Particularly not training people in "ISO9000" or "How To Be Audited."

Training should be carried out to develop awareness and skills, and should be provided as part of an action plan to address a competency needing to be improved[11].

Interestingly, the clue to the sequence of implementation is in the title of the requirement.

Competence is defined in the vocabulary document ISO9000 as:

"The demonstrated ability to apply skills and knowledge"[12]

So, why can't the grandfathering card be played? After all, some people have been working at their jobs a long time! Because there's a need to determine and demonstrate competence for personnel who perform work which affects Quality. Although there are many theories surrounding how

---

[11] Adapted from ISO 9001:2008, section 6.2.2, page 6, with the permission of ANSI on behalf of ISO. © ISO 2013 – All rights reserved.

[12] This excerpt is taken from ISO 9000:2005, with the permission of ANSI on behalf of ISO. © ISO 2013 – All rights reserved.

adults learn, there are generally accepted phases which describe the progression:

- unconscious incompetence,
- conscious incompetence,
- conscious competence, and
- unconscious competence.

It would be easy to assume that an experienced person is someone who has reached unconscious competence, who could be grandfathered into their job. That would be potentially a missed opportunity, partly because these are not concrete phases.

An editorial article published in a British motorcycle enthusiasts' magazine, discussed the significant numbers of British motorcyclists who were being killed in road accidents. There were some interesting statistics, including that often no one else was involved, just the rider. Another was the average age of the riders, which was 50 years. So what was causing these riders to get into these fatal events? By analysis of the facts associated with each crash, it can be seen that changes had occurred and these had affected the riders' abilities – from eyesight acuity and speed of reaction to the power and agility of the motorcycles being ridden. All had an effect on competence, compared to their early years of training or experience. The final analysis of these events revealed that just because someone attains phase 4, it doesn't mean they stay there!

Although it may seem important that the organization communicate that it is using ISO9001 to develop a formalized approach to its process controls, too much reference to "ISO" can be counterproductive. In an effective implementation, most of the emphasis goes on ensuring that people are aware of the fundamentals of the organization's quality management system, and of why these are important to them in getting their work done, for regulatory compliance and to ensure the customers' needs are met, etc.

There are many ways to communicate the reasons why ISO9001 has been selected, why the organization has chosen to formally define and implement a quality management system and ultimately become certified by a third party, why training isn't likely to be required for 99% of personnel. Far better to describe these through the use of "brown bag" lunch meetings, newsletters, staff meetings, and other events – and no records are required!

Infrastructure (subsection 6.3)

Try as we might, it's a practical impossibility to produce a good-quality product without the appropriate equipment, power, computers, information, and so on. The infrastructure requirements which are needed to ensure product conformity should be identified and provided by the organization, including such items as process equipment, buildings, utilities, workspace, and services.

Work environment (subsection 6.4)

In step with the above, the work environment can also have an effect on product conformity. There is a note underneath the ISO9001 requirement which defines typical environmental factors, such as noise, humidity, lighting, or weather. Others might include temperature, air quality, dust, etc. These factors have to be managed.

Product realization (Section 7)

Weird name, "product realization"! What it is referring to is the various processes which take the input of customer requirements and transform them into deliverable products and services (for the most part).

Planning of product realization (subsection 7.1)

Myth alert!

Our processes are simple; we don't need to plan.

We don't have time to plan our processes.

If we revisit the very first requirement of ISO9001, we can see that it's the organization's responsibility to determine the processes needed for the quality management system and their sequence and interaction. In smaller organizations this almost becomes the plan for the way that customer inputs are turned into deliverable products.

Other organizations – often those which perform machining, stamping, or assembly operations – use a simple job traveler or operations sheet which defines each step or operation, in sequence, which transforms the raw material into the completed product. It's also common for quality control checks to be required to verify that the product is correct to specification.

Sometimes more complex products, or contractual arrangements with a customer, dictate a more detailed document which may be called a "quality plan."

Customer-related processes (subsection 7.2)
Myth alert!

We call this "contract review."

Our customer only gives us verbal instructions; there is no "contract."

This is just about having a process for customer orders.

"Quality," in all respects, begins with determining what the product being offered (for sale) is and what the customer wants, with the intent of coming to some form of agreement to supply! These requirements fit the convention of quoting, processing orders, and changing orders, in whatever form these take.

Two distinct requirements are stated:

Determination of requirements related to the product (clause 7.2.1)

The requirement touches on four basic factors:

- determining customer requirements;
- determining requirements not stated by the customer but still necessary for use, where known;
- statutory and regulatory requirements; and
- anything else thought necessary by the organization[13].

Often, with innovative new product designs, these factors are determined during development, through some type of

---

[13] Adapted from ISO 9001:2008, section 7.2.1, page 7, with the permission of ANSI on behalf of ISO. © ISO 2013 – All rights reserved.

market/customer research, sometime before sales or agreement to supply the customer occurs. Indeed, a note at the end of the requirement indicates that the review may be made of product catalogs or advertising materials, website descriptions etc.

Review of requirements related to the product (clause 7.2.2)

Before any commitment is made to supply customers, a review must be undertaken. Once again, the review isn't simply a case of checking a customer order for the stated requirements – although that is important – but takes place on receipt of a customer request for quote (RFQ/RFP), on receipt of an order, and when changed orders are received by the organization.

The purpose of the review is to ensure that

- the product requirements are defined;
- the order and any previous offer (proposal, for example) are compared and differences resolved; and
- the organization has the ability to meet the stated requirements[14].

Records of the reviews and actions which arise are required to be maintained.

It's understood that there are situations where organizations that take orders from customers verbally (telesales), and without any documents being used, cannot reasonably perform a review. In such cases, confirmation of the

---

[14] Adapted from ISO 9001:2008, section 7.2.2, page 7, with the permission of ANSI on behalf of ISO. © ISO 2013 – All rights reserved.

customer's requirements is necessary. This can take the form of an order confirmation (e-mail, fax, etc.) or by verbally confirming with the customer (which is often recorded).

Design and development process (subsection 7.3)
Myth alert!

The product design and development process can't be measured.

Design engineers need the freedom to be creative.

Following "ISO" will slow the process down.

The product design and development process requirements of ISO9001 are fairly simple and intended to ensure that the organization only designs products that can be produced, installed, and serviced without problems and that are safe for use – and that customers get what they really want!

Often known as the new product development (NPD or similar) cycle, the ISO9001 requirements form a basic framework for controls which should enable, not reduce, creativity. The ISO9001, section 7.3, requirements cover:

- design development planning,
- design and development inputs,
- design and development outputs,
- design and development reviews,
- design and development verification,
- design and development validation, and
- control of design and development changes.

Experience shows that there are a few key indications of how effective the design and development process is, and these are easily identified and measured:

- the number of post-release design changes (not resulting from customer inputs),
- the "on-time" and "on-budget" performance, and
- the retention of engineering staff.

The last point is directly linked to the first; however, it is rarely seen in this light. Let's take a look at a typical new product design engineering process, so we can understand the indicators of an ineffective process:

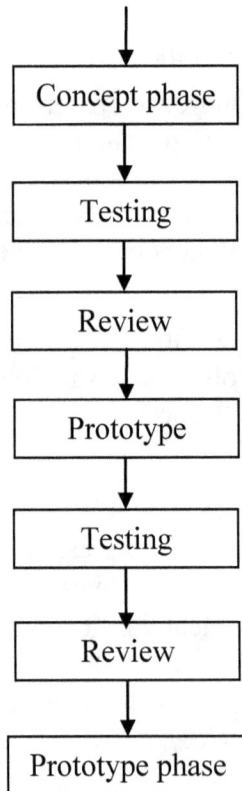

```
          │
          ▼
┌─────────────────────┐
│   Concept phase     │
└─────────────────────┘
          │
          ▼
┌─────────────────────┐
│      Testing        │
└─────────────────────┘
          │
          ▼
┌─────────────────────┐
│      Review         │
└─────────────────────┘
          │
          ▼
┌─────────────────────┐
│     Prototype       │
└─────────────────────┘
          │
          ▼
┌─────────────────────┐
│      Testing        │
└─────────────────────┘
          │
          ▼
┌─────────────────────┐
│      Review         │
└─────────────────────┘
          │
          ▼
┌─────────────────────┐
│   Prototype phase   │
└─────────────────────┘
```

```
                        │
                        ▼
          ┌──────────────────────────┐
          │         Testing          │
          └──────────────────────────┘
                        │
                        ▼
          ┌──────────────────────────┐
          │         Review           │
          └──────────────────────────┘
```

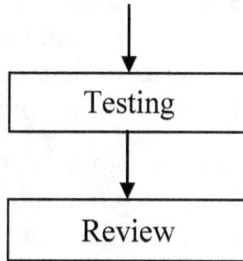

A robust product design and development process should release a set of product specifications to manufacturing and (possibly) servicing organizations (whether performed internally by the organization or by its suppliers) which can be produced within the capabilities of those manufacturing and maintenance processes. The product of the design engineering process is not unlike the product of the manufacturing process(es) in that it shouldn't be necessary to "touch it twice" to make it (to specification) and it must be delivered on time and at the planned cost. Therefore, three key indicators of the effectiveness of the new product design and development process are:

- delivery to schedule,
- development cost on budget, and
- post-release changes.

This last point has been well studied, because the cost of changing a product design after release to production is a significant factor. It's suggested that a fault found in the design phases might cost x to rectify, would cost 10 times that value ($10x$) if detected in production, and would cost 100 times more ($100x$) if found while in customers' hands.

Often, it's necessary to process product design changes once the product has been released to be produced, either by the organization's own manufacturing function or by their suppliers. Experience has shown that often such changes are given to those design engineers who have recently qualified and been recruited, much to their chagrin.

Newly qualified product design engineers probably have the impression that, once they take a position in industry, designing new products is exciting, that they will get to use their expertise, bringing innovative and creative ideas to light. In reality, they may be assigned (initially) to work on existing product design changes. This presents two challenges for these engineers:

- it's often very difficult to reengineer a product, and
- they are often not sufficiently experienced to reengineer products and this can be demotivating.

A study of major Japanese new product introductions showed that the majority of changes to product designs occurred before release into the marketplace, and that few changes were undertaken after that. The results? Improved customer perception, lower warranty costs, happier service staff and improved market image, to name a few.

Purchasing (subsection 7.4)

Myth alert!

An approved supplier list is required.

Suppliers need to be audited.

All products have to be inspected upon receipt.

All suppliers should be ISO certified.

Quality in the supply chain of any organization is as important to the quality of the end product/service as anything the organization does itself. As more and more organizations perform fewer and fewer of the processes which create products, the importance of suppliers increases.

Despite many organizations having complex (and often complicated) procurement methods and procedures, ISO9001 really states a fairly simple set of requirements:

- Make sure what you buy is what you want.
- Make sure you control the supplier and their product based on the effects on final product.
- Evaluate, and select suppliers based on capability and then reevaluate suppliers to established criteria.

It's not so different in many ways to the way, as individuals, we buy from our local grocery store! When we buy fruit and vegetables, we often perform some type of inspection to check ripeness and damage etc. It may, however, be difficult to tell if potatoes have frost damage, which may only be revealed when pealing them.

Dairy products are checked for their "best-by" date, seen stamped on the container, and eggs may be quickly checked for damaged shells.

Prepackaged foods like canned soups, breakfast cereals, and cookies are quickly checked for damage to the packaging and that's all.

Each type of grocery product is given controls based on the quality requirements, and often our experience of quality issues plays a part in the controls we employ – no one wants to arrive home only to discover that their eggs are all broken, the milk is three days past its expiry date, or some peaches are badly bruised or overripe and rotting.

Purchasing information (clause 7.4.2)

Suppliers must be given clear and unequivocal information on what is wanted by the organization. Some suppliers may supply items from catalogs for which they maintain the specification, and in this case it is appropriate to use this specification.

Unfortunately, since there are cases where what is being purchased may come from specialist suppliers, it can be difficult to create a specification which is meaningful. This also includes outsourced processes, where the organization's products are sent to a supplier who does additional value-adding work to the product. As mentioned in the first part of this chapter, it is frequently the case that the work done by the suppliers is not well understood by the purchasing organization (hence its being outsourced) and is subsequently difficult to verify to a specification. Consider a component which is sent to a supplier outside the organization for plating or painting: how does the organization ensure that the supplier did a quality job and the coating won't fall off later? It can't be checked (both are so called "special" processes – see 7.5.2).

Provision may be made under this requirement to detail a number of controls and deliverables to be implemented by the supplier in ensuring product quality, including product, process, procedure, and/or equipment approvals. Personnel qualifications may also be specified, as can quality system requirements. In industries where welding is critical to product performance and safety, it is common to specify the qualifications of the actual people doing the welding – say, to ASME Section IX, or AWS D1.1 in the USA. In addition, the organization may contract with a welding engineer to review and approve the welding procedure specifications (WPS) and also require delivery of the procedure qualification record (PQR) following testing and inspection of the weld(s).

Verification of purchased product (clause 7.4.3)

As previously stated, the organization has to ensure that what it buys is correct to the specification. This may be done by waiting until the product is received at the organization, or at the supplier's premises. We may find it perfectly acceptable – and practical – to buy those potatoes, knowing that it's not until we peel them that we find out how good they are.

The clause goes on to describe that where the organization – or its customer(s) – intend to visit the supplier to perform verification of the product, the arrangements are to be specified in the purchasing information to that supplier. No one wants any surprises!

This last requirement isn't as much the subject of mythology as it is just plain misunderstood! It is a legacy from the antecedents of the ISO standard – the military supplier quality requirements. When significant purchases of military equipment are made, it would be highly risky to wait until delivery of a ship's engine to determine if it was built and performed to the agreed specification. Therefore, the engine supplier is required to make arrangements at various stages of build and final test. Indeed, the ship itself goes through extensive sea trials before being handed over to the Navy!

Production and service provision (subsection 7.5)

Control of production and service provision (clause 7.5.1)
Myth alert!

All processes need work instructions.

You should document your process so that if no one turns up for work, anyone can do the job.

For many organizations, this requirement relates to the heart of the business process of getting product to customers. Although the requirements are seemingly simple, what is often missed is the concept behind controlling the production and service processes, which is to have "standard work." That is to say, each activity can be replicated over and over again, without deviation leading to nonconforming product or process. An example which is generally well known by almost everyone is a famous fast-food restaurant chain, which has produced the same basic hamburger, serving billions by following the same processes and methods.

There are six basic requirements for the control of production and (where applicable) service:

- the availability of information about the product's characteristics;
- the availability of work instructions, as necessary;
- the use of suitable equipment;

- the availability and use of equipment for monitoring and measurement;
- the implementation of monitoring and measurement activities; and
- the implementation of product release, delivery, and post-delivery activities[15].

It is true to say that any one requirement should not be considered without reference to at least one other. This particular requirement of ISO9001 specifically requires consideration of a significant number of others. Take the requirement for the "availability of work instructions," for example. The addition of the words "as necessary" might be considered "weasel words." They allow us to "weasel" out of having to comply! So, is it justifiable to not employ works instructions? A clue is in understanding the relationship to other ISO9001 requirements which have influence on subsection 7.5.

As noted previously, the people who are responsible for the control of a process may be assigned based upon competence (clause 6.2.2). Information about the product may be provided through the use of blueprints, drawings, and specifications for the product, which all fall under the document control procedure (clause 4.5.2). While operating the process, it's normal to expect product characteristics to be measured or monitored (clause 8.2.3) and process parameters to be measured or monitored (clause 8.2.4). These results become records (clause 4.2.4). The equipment used to monitor and/or measure will also be calibrated or verified to ensure accuracy of the recorded results

---

[15] Adapted from ISO 9001:2008, section 7.5.1, page 10, with the permission of ANSI on behalf of ISO. © ISO 2013 – All rights reserved.

(subsection 7.6). In addition, the process equipment will be required to be maintained (subsection 6.3)

It will be seen that a network of requirements is, in practical terms, what is needed to ensure process control – for which one option might be to give the operator an instruction!

Validation of process for production and service provision (clause 7.5.2)

Myth alert!

This requirement doesn't apply to us – we've been doing these processes for years! They aren't "special."

We test a sample to tell us if the others meet the specification.

For some processes, the quality of the resulting output cannot be reasonably tested or inspected. This requirement of ISO9001 defines what the organization must do to ensure that an acceptable product is produced. For many, actually determining what processes fall into this category can be a challenge. In previous versions of ISO9001, these processes were known as "special processes," but even this didn't adequately describe them – for such processes are quite common for organizations who perform them and are not, therefore, considered "special"!

So what processes fall into the category of needing validation? Some typical examples include;

- gluing and bonding;
- welding, brazing, and soldering;
- heat treatment and sterilization; and
- plating, painting, and coating.

2: ISO9001 and Its Myths

As anyone who has attempted to repair a broken item at home (let's say a china vase), we know that following the instructions on the glue packaging is of vital importance to a good result. We're told to "clean the mating parts," often with some proprietary degreaser, spirits, or cleaner. The mating surfaces being joined might be roughened with sandpaper to provide a "key" for the adhesive. Of course, the temperature may also need to be between some values which represent a typical summer's day.

Having prepared the surfaces, it may be necessary to coat the mating parts with the adhesive and allow it to cure (at least partially, when it's less tacky and somewhat dry to the touch). At this point, the mating parts may be required to be brought together and held, often under some pressure for a defined amount of time. We also know it's tempting to test our newly created join in the precious vase to see if it's fixed! We might try to actually break the joint! How quickly our surprise turns to frustration when the part gives way and we have to start over! Oh, the temptation to check the joint once more …

Performing a test or similar on a sample from the process is a way to reduce the risk of failure in the other products being processed. This is still fraught with potential problems and can be wasteful too. We're left with some imponderable questions about how representative the sample is. How is that validated? By the time steps have been taken to check that, the requirements for process validation have been met, so there's now no point in sacrificing a sample!

Identification and traceability (clause 7.5.3)
Myth alert!

Everything needs a part number.

All our products have to be traceable to the raw materials.

This is a requirement of ISO9001 which is full of "weasel words"! It starts "Where appropriate ..." This leaves the organization some latitude to decide what's appropriate. It should be clear that – especially when products look similar – some form of identification may be necessary to prevent problems caused by mixing them. The methods used for identification can be quite diverse; identification on the part, location, or container labeling are common options. It's not required when there's sufficient difference for people to be clear about which product is which.

The second requirement – the status of the product in regard to inspection or testing – is clearer. The organization "shall." The reason? Because untested or uninspected product shouldn't be allowed to progress and only known "good" product should make it to the point of being shipped to a customer! Or, retesting or reinspecting a product is expensive, causes delays, and adds no value. Everyone knows how frustrating it can be changing batteries and getting the old and the new ones mixed up! All for the sake of segregating, tagging, or marking them in some way to ensure they don't get mixed.

Traceability is one of those areas where, again, "weasel words" are used. Simply put, traceability allows the organization to limit its exposure to the risk of quality problems as far up and down the supply chain as is practical or necessary. Often driven by regulatory requirements, unique identification is normally applied to product and/or its associated packaging

Customer property (clause 7.5.4)

Customer property is anything the customer gives the organization which is used in the deliverables to that customer. When dealing with customer property, it's normal to invoke a number of other requirements of ISO9001, to handle the processes which deliver the property back to the customer. A traditional application of this would be a car taken into a dealership or garage for servicing or repairs. The car represents "customer property" and, as such, it's the dealership's responsibility to:

- Take care of it to ensure no damage is done to the car while it's in their custody, either by their staff or while parked on their premises (vandalized) and that they can locate it when it's picked up by the owner.

- Report anything found during the work which may affect the ability of the work to be carried out. For example, if the car had a nonfunctioning parking brake, the dealership would probably not want to be responsible, since the car may run away down a slope, when "parked."

Organizations who repair products – perhaps in addition to designing and manufacturing them – deal with customer property as a core process. It's not unusual for product to "appear" on the receiving dock with minimal or even zero instruction on what the customer required to be done, or without even a purchase order. In such a case, it's up to the organization to determine what work is required and report this to the customer, as the basis for establishing agreement on cost and return or delivery.

Preservation of product (clause 7.5.5)

Happily, there don't appear to be many common myths with this requirement. It's generally understood that it's important to take care of everything from incoming raw materials, through work in progress, to final product. Preservation can mean not only preservation of the product, keeping it to specification and free of damage caused by corrosion, physical damage, etc., but also preservation of quantity.

Keeping an accurate inventory – by quantity and location (where it's applicable to the product) – is key to ensuring customer requirements can be met and production isn't halted awaiting the correct quantity of components.

The requirement extends to the "constituent parts," so any components should be included in the controls.

Control of monitoring and measuring equipment (subsection 7.6)

This requirement of ISO9001 (subsection 7.6) is possibly one of the most confused, and somewhat abused, in addition to being a "target-rich environment" for auditors. Despite this, the science of calibration starts out being a very simple concept applied to an item of measurement equipment. Certainly, if done correctly, there are mathematics and related activities which can become major potholes into which the uninitiated may fall headlong. Further, even some of those experienced people who have worked in the functions usually responsible for the calibration and maintenance of equipment are often unclear about much of the application of the science. In spite of the potential for rocks in the road, if we recognize the multiple myths and navigate around them, the road to properly controlled equipment can be very smooth.

Myth alert!

Common myths surrounding the control of measuring equipment include:

All equipment has to have a "calibration" sticker on it to show the date calibrated, recall date, etc.

Calibration must occur to a time base (annually, semiannually etc.).

All measuring equipment must be calibrated.

Calibration includes adjustment to correct the reading.

Uncalibrated equipment should be labeled "for reference only."

Only equipment used for final inspection needs calibration.

Calibration is expensive.

Trusting measurement

To understand the need for calibration, it's necessary to understand some basics of the science of metrological calibration. In its most basic form, the act of calibration is a comparison to determine how much a measurement reading or result deviates from the corresponding finite value. Many of us take measurements which affect our lives, from stepping on a weighing scale to discover our body weight, to driving a car along the road. Both items of equipment include an indicator of some kind. The weighing scale and the speedometer look a little like a clock or have a digital "display." When we step on the scales or drive along the road at, say, 30 mph (or 45 kph) we assume that the value indicated is true and accurate.

In reality, each time we take a look at the measurement of our weight or the car's speed, we are not looking at a true value being indicated. A car speedometer is allowed, by regulation (at least in the USA and in Europe) to deviate by up to 10% of the indicated value. This means the car might, in reality, be moving down the road at between 27 mph and 33 mph. Have you noticed that a typical car speedometer has divisions of 5 mph/kph?

As the owner of the scales or car, we usually have no access to any significant type of adjustment which affects the value displayed, so we might ask just how the manufacturer of the weighing scales or the car speedometer knew what value to make the indicator display. During the manufacture of these indicators, care is taken to ensure the units are each made alike. Of course, there are minute variations in the relative sizes of all the component parts. When the total part variation is considered, it can affect the value being displayed in a positive manner (the value is increased) or negatively (the value is decreased) – but by how much? And, more importantly, does it matter?

Accuracy and precision

Keeping with the examples of the weighing scales and the car speedometer, the accuracy of the value indicated is important in one case – because of the need to comply with regulations – the other is a "ballpark" figure. Hence, the speedometer has to be calibrated before it's installed in the car and the amount it is "off" the actual speed is then known. Furthermore, most traffic police use speed detection devices, like radar, which have greater accuracy and precision than the regular speedometer. The police measure your speed with less error, so that if you receive a speeding ticket, it's highly likely that the value stated on the ticket is accurate to within a few percent of the actual speed the car was moving.

Measurement, analysis, and improvement (Section 8)

General (subsection 8.1)

Having established a quality management system, and set objectives and controls in place for the various processes of the organization, it would be appropriate to perform some measurements to see how the QMS is performing. As part of the "Plan, Do, Check, Act" cycle, this requirement of ISO9001 can be considered to be "checking" as a precursor to the "Act" of improvement.

The general requirements specify three basic areas of importance to be planned and implemented as measurement/monitoring activities, leading to the identification of action needed for improvement:

- product – to demonstrate conformity,

- quality management system – to demonstrate conformity, and
- continual improvement – of the effectiveness of the QMS.

Included in the General section of the Standard are pointers to the methods of measuring to be identified, including the use of statistical techniques and to what extent they are used.

Monitoring and measurement (subsection 8.2)

For some reason, this requirement causes a lot of confusion. It may be, in part, because there's so much about an organization's processes which can be monitored and measured. Confusion also stems from the monitoring, compared to measuring, of processes, which will be explained below.

When considering what has to be considered, customers come first!

Customer satisfaction (clause 8.2.1)

Myth alert!

Customer satisfaction surveys have to be sent.

Customers are satisfied if they order more and don't make complaints.

We send a survey and customers don't send them back, so we're doing OK.

Customers keep buying from us, so we don't need to survey them.

It may seem obvious, but if you're going to "do" quality, it makes sense to ask the customer(s) of the organization if quality has been achieved. Careful review of this requirement shows that it's not customer satisfaction per se that is to be measured, but information relating to the customer's perception of whether their requirements had been met.

As we know, it's often the unspoken comments of a customer which tell the most about their satisfaction. Having dined at a restaurant it's not unusual to be asked, as you leave, "Was everything alright?" Replies are sometimes a simple "yes, fine," when in fact there was more that could have been said to have given a fuller picture of the customer's satisfaction.

This requirement is strongly linked to the quality policy and objectives which the organization's management set. On that topic, we discussed whether it was truly intended to "meet and exceed customer expectations," especially since expectations aren't always communicated. If an organization is truly interested in customer feedback, it will institute some person-to-person discussions and practice some "active listening" when customers (or their representatives) have been asked about their satisfaction or about what could be improved.

Naturally, having been asked and also having given the organization their feedback, it's highly likely that customers will expect the organization to take action!

Internal audits (clause 8.2.2)

Myth alert!

You have to cover all the ISO clauses, once a year/every three years.

One internal audit per year is sufficient.

Audit findings should be graded as "major," "minor" or "OFI" (opportunity for improvement).

An audit calendar is required to show what's going to be audited in the coming year.

Auditors must be chosen from a different department than the one they audit, so they are "independent."

Someone must be the "lead auditor."

Auditors must have a checklist.

The audit process has to be measured.

Internal audits are one of the requirements of ISO9001 which are not "institutional" to the way an organization works, and possibly many myths plague their implementation. Some form of audit is a familiar practice to most organizations, although it is often from a financial perspective or, as has been discussed earlier (in the Introduction) through being evaluated by a customer.

Since internal audits are required and, as we know, external "certification" of compliance is not necessarily a goal for an organization which utilizes ISO9001, what is the purpose of including these audits in the Standard?

Naturally, verification of compliance of the organization's quality management system with the ISO9001 Standard is a given. The organization must determine that its quality management system meets the requirements of ISO9001, as long as management requires it. But that's often where the use of internal audits starts and stops. Let's see why.

It is not uncommon for internal audits to start with sending someone to an auditor training course. This is a great way to get started on the basics of an actual audit. The agenda for most courses will cover the fundamentals of planning, preparing, conducting, and reporting audits. A typical internal auditors' training course can be two or three days in duration (16–24 hours), or the option exists to take a "lead auditor" course which is 32–36 hours in duration. Many are even accredited by a body such as the RABQSA or IRCA, which means they are recognized as being designed and delivered following the requirements of the International Standard on Management Systems Auditing, ISO19011.

Once trained, it's tempting for the new auditor to sit down and start creating a checklist or an audit schedule or calendar which shows what is intended to be audited in the coming year. An organization cannot be successful if it requires certification of its quality management system without having implemented internal audits.

In the run-up to the certification of the organization's management system, the focus of the internal audit program is to ensure that no significant issues are found during the certification body (CB) audit. In the chapter on certification, the CB auditor will confirm that audits have been implemented at the "Stage 1" audit; therefore, it becomes a key success factor in the lead-up to that phase.

Once found in compliance and successfully certified, the internal audits should be scheduled to ensure that their focus changes to bring another value to the organization, rather than simple compliance with the ISO standard. To do this, the scheduling of audits must be changed from a "push" system, where audits are scheduled according to an annual "calendar," to a "pull" or demand system, where they are scheduled according to management's needs for the business.

The ISO9001 requirements for internal audits (clause 8.2.2) give us a clue to what might be considered when establishing an audit schedule (also called an audit program). It states (in part): "An audit program shall be established taking into consideration the status and importance of the processes to be audited." This requirement gives us some opportunities to consider that all processes are not created equally – that some might need to be audited as a priority and possibly more frequently than others. So, how can an audit program meet these requirements and add value to an organization?

Let's consider what's meant by the status of the processes of the quality management system. Status might include something being new and/or changed, performing below or above expectations. Having something new or changed in relation to the following is frequently also associated with causing problems:

- customers and requirements;
- suppliers;
- technology;
- regulations;
- process requirements; and
- materials, equipment, etc.

The above are normally considered to be risks to the business. In fact, how many of us have heard that it's not a good idea to buy a new model car or similar until after it's been on the market for a year – to "work the bugs out"?

Likewise, a process not performing to expectations, causing scrap, rework, and downtime – in fact any kind of waste – is also a risk. Got a process which exceeded its goal? Better find out why! Once the reason has been discovered, it could be used to improve other, similar processes.

In some cases, these are planned situations – including new and changed aspects. Some, like poor performance, are unplanned. What could be done to give a priority to an audit of a new, changed, or poorly performing process? This is where the importance of that process must be considered. We have to ask ourselves: is the process important to meeting

- customer needs and expectations, and/or
- regulatory compliance, and/or
- costs?

The importance of the process to the customer or other aspects of business can be considered as the impact of that process on the business.

It is common for internal audit programs to be developed on an annual calendar which predicts which aspects of the quality management system are going to be audited. Often the objectives for developing the schedule are to ensure that the entire system is audited in that year, or to ensure that all ISO requirements are covered, etc. However, since there is no requirement to perform audits in that way, these internal audits often miss critical processes when they become an issue.

In one company, interns were recruited to cover an assembly line during summer vacation time, which resulted in almost predictable problems with product quality. The internal audit schedule forced the auditors to focus on features of the management system which were rarely problematic – instead of taking a critical look at the training and competence of the new people.

Imagine the hapless line supervisor having to act like a mother hen to his or her new operators while attempting to answer auditor's questions about product labeling, document control, etc. Had the audit schedule directed the auditors to the training process when it was being implemented, it's highly likely to have diagnosed the problem with the process and attracted management's support for corrective actions.

An annualized calendar forces audits either of processes which are not a high priority, or before/after any problems transpire, instead of helping to identify what contemporary actions need to be taken to improve things. No wonder, then, that in many organizations, the internal audit program is not well supported! Internal audits should be scheduled using current process performance data, feedback from customers, etc., to ensure that auditors are focused on what is on management's "radar screen."

Internal audit management programs, scheduled according to risk and impact, can help usher in a new era, synonymous with risk assessment and continual improvements, rather than something done simply for compliance. Furthermore, the role of auditor becomes elevated in status similar to that of a SixSigma/Kaizen practitioner. Improving the internal audit program in this manner will help to ensure a domino effect on corrective actions and on management reviews of the quality management system as a whole.

The selection of people to be auditors is made on the basis of finding someone who is "independent" of the process being audited. Someone from manufacturing can audit purchasing, someone from sales gets to audit design engineering, someone from customer service can audit sales.

Although it is important to maintain impartiality of auditors, this is one technique employed by certification bodies which can be successfully mimicked. Certification body auditors are qualified by industrial sector codes (for example, EAC/NAICS, etc.), which means, in broad terms, that they have experience in that sector of industry. So, why not consider that internal auditors should have a passable knowledge of the process/function they are assigned to audit?

Although many will say internal audits are a process and should be measured, experience shows that there are no meaningful metrics to show they are effective. Some have suggested that the number of nonconformities or conformities to an audit schedule is a measure of internal audits; however, and in all reality, what is a "good number" of audit nonconformities? One, or 10? Indeed, since the requirement appears in the measurement and monitoring section of the ISO9001 Standard, they are themselves an activity of monitoring or measuring the processes of the Quality Management System. The best we can hope for is that the organization's management takes an active, leadership role in deciding what and where internal audits should be performed.

Monitoring and measurement of processes (clause 8.2.3)

Myth alert!

All our processes should be measured.

Some of our processes cannot be measured.

The measurement of processes is of significant importance when considering the effectiveness of the quality management system in meeting customers' requirements and also the stated quality objectives. The responsiveness, for example, of the process which provides a customer with a proposal or a sales quote, or for taking a field service call, is typical of measurements important to a customer.

Monitoring and measuring are, for those who own and/or drive a vehicle of some sort, something we do each time we take a trip. Car owners and drivers use the information displayed on the dashboard of their car to help them during a journey. Let's review what information is provided to get a sense of monitoring and measuring and why it's important:

Speed

The speed of the car is something which should be monitored and measured. Why? Because there are laws that drivers have to comply with on any given journey. If you don't pay attention, you'll be fined!

Fuel

It's normal to monitor the amount of fuel used during a journey to ensure it doesn't run out! We may be interested in the actual amount of fuel used, to ensure that the vehicle is obtaining a certain specific fuel economy, is running properly, etc.

Engine condition

Either through the use of the "rev counter," which indicates the speed the engine is turning, or from the coolant temperature, we can tell something of the running conditions of the engine – although we never actually need to know the actual values (say, of the coolant, which is normally at about 74 degrees Fahrenheit). Knowing the actual value isn't important, which is why most temperature gauges show only "C" for cold, "N" for normal or "H" for hot, or by using blue for cold, green for normal and red for hot. The actual temperature values at these positions on the gauge aren't of interest per se, but the position in the range is, for monitoring purposes.

There are many other values which can be used to indicate that an engine is performing correctly – oil pressure, inlet vacuum, oil temperature, for example. Other values could be taken from the ancillary equipment on the engine, like the voltage and current from the generator, battery condition, ignition dwell angle, etc.

Indeed, the modern Formula 1 Grand Prix racing car has thousands of measurements made every second, of hundreds of parameters, to give the many engineers responsible a total picture of just about everything the car is doing in response to the driver's control inputs. But then, the objectives – and costs – of running a Formula 1 GP racing car are totally different to those of driving an everyday road car.

We can also see that if all these values were to be presented to us as the driver, we'd soon be overwhelmed. Indeed, some of these values are only of use if we're diagnosing a problem.

By keeping an eye on the basics, on speed and fuel, most drivers can ensure they will complete their journey objectives: on time (speed over distance), on cost (fuel used), and without penalty (speeding ticket).

If we take this analogy in the context of the organization's quality management system, it can be seen that when it comes to measurement and monitoring processes, it is important to consider any statutory or regulatory requirements which may affect the processes. In lieu of these, the customer is next in line when it comes to identifying what's important to measure/monitor. After this, the need for better control and improvement – which may include the use of some form of statistical control technique, would be chosen by the organization.

Monitoring and measuring of product (clause 8.2.4)

Myth alert!

This only applies to final product.

If the product is to be delivered into a regulated market, it will be of importance to be able to monitor and measure product characteristics. Often based on functionality and the capability of the process, product features and characteristics may be required to be checked by an operator, sampled by the QC function, or submitted to a laboratory for analysis. In the aerospace industry it is common for a supplier to submit a "first article" with a complete "layout" or physical inspection of a machined component.

Identification of key features and characteristics of products is very common in many markets – automotive customers, for example, will identify classes of features on their product specifications, in terms of product safety (handed down from them through the supply chain), appearance, and fit/function, for example. These characteristics are called out in a "control plan," which details what about the product should be checked and when, often in process, since waiting for the final, finished product to be evaluated would have significant knock-on effects if it were found to be nonconforming!

Control of nonconforming product (subsection 8.3)

One of the reasons for monitoring and measuring product is to detect any nonconformities (rejects) before they escape the organization and get to the customer. In regulated industries, this is viewed as very important, to ensure that defective product cannot be shipped, and therefore used. When taking the basic controls into consideration, there are many myths which surround the controls for nonconforming product.

Myth alert!

Nonconforming product must be moved out of the production process, segregated, and/or locked away.

Everything found nonconforming is "scrap."

You have to have a Material Review Board to decide what to do with nonconforming product.

Nonconforming product must be marked as such.

A nonconformance report has to be completed for each defect type.

We can allow the people who find the defects to decide what to do with the nonconforming product.

The full requirements for controls for nonconforming product are, experience shows, overlooked. The simple(r) aspects, such as tagging or disposition, are frequently given more emphasis than those which help to identify what caused the nonconformity and how it can be effectively processed subsequently. In particular, the decision on the disposition of nonconforming product has to be given to the "relevant authority" and potentially by the customer.

In addition to the quality control of products actually made by the organization, another feature of monitoring and measurement would be the products and materials sent to the organization by suppliers.

Analysis of data (subsection 8.4)

Myth alert!

We are ISO registered; our QMS must be suitable and effective.

We keep all our records, just like our procedure says.

The measurement of product and process should generate quite a lot of data which should be used to give the organization's management an indication of the achievement of the established objectives – or at least of progress toward them. The creation of records of monitoring and measurement isn't there just to comply with the Standard; the data they contain can tell important stories not only about the achievement of conformance to specification for products, or about the performance to objectives of processes, but also about activities where savings can be made. An example often overlooked is that of calibration of measuring equipment.

When establishing a frequency for recall and recalibration (often an expensive and disruptive event), the manufacturer of the equipment may advise a yearly schedule. Experience shows that if the "as-found" calibration data is reviewed, the equipment may be sufficiently robust (or had careful/infrequent use) such that a recall may not be required (if conditions aren't changed) for two or three years! It can be seen that, without risk, significant savings can be made over five to 10 years, by not "over-calibrating."

By considering this requirement with the control of records, an effective use of data as input to decision making is of vital importance to the organization in guiding the next activities of the Standard.

Improvement (subsection 8.5)

Myth alert!

There's a difference between "continuous" and "continual" improvement.

It's necessary to have a procedure for continuous improvement.

Corrective action needs root-cause analysis.

Corrective action must be taken on all nonconformities.

Corrective actions must lead to preventive action.

We retrained the operator.

Continual improvement (clause 8.5.1)

One requirement missing from all previous versions of ISO9001 was the need for the organization to make any form of improvement! The use of popular improvement tools such as SixSigma or Lean, which came to prominence during the 1990s, made this apparent "hole" in ISO9001 seem larger and more obvious. Consideration of the eight management principles includes continuous improvement.

Although these words may not have the exact same meaning in the English language, for the purposes of applying the ISO Standard, they can be considered to be simply different approaches to achieving the same objective.

The approach the organization adopts in implementing any improvement should be appropriate to its needs. For example, if we consider the SixSigma methodology, this can be considered a project or so called "breakthrough" type of improvement. People trained and qualified as either "black belts" or "green belts" comprise a team of improvement specialists. They participate in a project, following the "DMAIC" steps (define, measure, analyze, improve, control) or similar. Implementation of SixSigma projects tends to take time to carry out, sometimes running into months, with periods of no improvement, often while the next opportunities are identified. As a result, a typical graph of improvements (measured in $) over time, might look like this step chart:

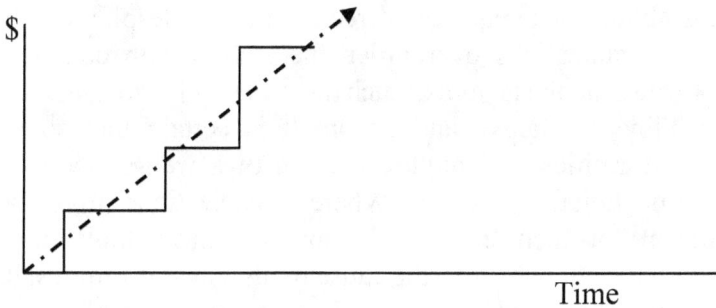

Improvements can also come from small-scale activities, such as the "Kaizen" events normally associated with the famous "Toyota Production System" or "Lean" manufacturing techniques. Kaizen is a Japanese word meaning "change for the better," and the techniques focus on the reduction of waste. Improvement events are often characterized by small-scale events of one or two days' duration which happen regularly – often weekly or monthly. They can be represented by an (almost) straight line. We can see that the net result is the same, in terms of savings over a similar time period.

Since any improvement activities are going to need resources – time, people, and money – then it's highly likely that the organization's management will have to approve those activities. Therefore, this requirement is a key aspect of the management review of the quality management system.

Corrective action (clause 8.5.2)

Corrective action is one of the least well practiced requirements of ISO9001, although many organizations believe they implement it effectively. The Standard is very clear about what requires corrective action, despite which many organizations overburden themselves with complex root-cause analysis tools, such as Kepner-Tragoe, 8D (8 Discipline or Steps), and so on. It is certain that some nonconformities will require the use of a well-structured, multifunctional approach. Where organizations make a problem for themselves is in applying these tools in a heavy-handed manner to the cause of all nonconformities. It is also recognized that organizations who supply into regulated markets – such as those in pharmaceuticals, medical devices etc. – have no option, since the use of "CAPA" is required.

Simply applying the tool of Pareto analysis (also known as the 80/20 rule), we can see that there is a relationship within the data we must analyze from the records generated by the product and process monitoring and measurement activities. The use of Pareto analysis tells us that the so-called "vital few" causes of the majority of product defects (for example) could be the focus of the structured corrective action. This leaves us with the resources to treat the "trivial" many under the natural actions allowed under the control of nonconforming product – the "disposition" options.

Preventive action (clause 8.5.3)

This requirement for preventive action is, perhaps, the least well understood outside industries such as the automotive industry. Since ISO9001 doesn't identify how to do something to meet the requirements, the first item becomes a significant hurdle to starting the process.

This section of the preventive-action requirement should cause a documented procedure to identify the preventive-action tools and techniques an organization could deploy, which range from a simple "lessons learned" review – to be conducted on receiving a new customer order or when undertaking a new design – to a comprehensive analysis of process failures and risk rating (known as a potential failure modes effects and analysis, or PFMEA) which may lead to the creation of a product quality control plan for product/process.

The apparent confusion surrounding preventive-action identification and implementation is reportedly one of the inputs to the proposed ISO9001:2015 revisions, as it will be replaced by some form of "risk management."

CHAPTER 3: ISO9004 – THE GUIDELINE DOCUMENT

As part of the "family" of ISO9000 documents, ISO9004 has been published with the intention of providing guidelines to those organizations seeking to implement the requirements specified in ISO9001.

Somewhat paradoxically, ISO9004 is mentioned only once within the ISO9001 requirements and only in the Introduction – something which is rarely considered by implementers. This is borne out by the numbers of the individual volumes sold by ISO – at one point ISO9001 outsold ISO9004 by a factor of 10 to one!

For those who did obtain a copy of ISO9004, users may not have been entirely satisfied with their acquisition! Let's briefly review the previous and current versions, since they are applicable to the ISO9001 document which has remained substantially unchanged since the 2000 version was issued.

ISO9004:2000

This version of ISO9004 was entitled *Quality Management Systems: Guidelines for Performance Improvements.*

The related abstract from the ISO website states:

This International Standard provides guidelines beyond the requirements given in ISO9001 in order to consider both the effectiveness and efficiency of a quality management system, and consequently the potential for improvement of the performance of an organization. When compared to ISO9001, the objectives of customer satisfaction and product quality are extended to include the satisfaction of interested parties and the performance of the organization.

This International Standard is applicable to the processes of the organization and consequently the quality management principles on which it is based can be deployed throughout the organization. The focus of this International Standard is the achievement of ongoing improvement, measured through the satisfaction of customers and other interested parties.

This International Standard consists of guidance and recommendations and is not intended for certification, regulatory or contractual use, nor as a guide to the implementation of ISO9001.

The guidance was published with the intention of being consistent with ISO9001 – the term "consistent pair" being used to describe them both. The format of the ISO9004 document was closely aligned to the ISO9001 requirements, so that cross-referencing was made easier.

ISO9004 was printed with the ISO9001 requirements incorporated within it, the guidance texts being laid out in juxtaposition. The fact that ISO9001 requirements 4 to 8 were included verbatim may have irked a number of users – when they discovered they'd already paid for the ISO9001 document as a stand-alone.

Furthermore, ISO9004 was never intended to be a "how-to" guide for those who were new to the concepts and practice of quality management systems, which would be frustrating, especially when the title states "guidelines."

It did, however, provide some excellent descriptions for those who sought to improve the basics of the quality management system they had constructed, in particular concerning internal audit planning, for example.

ISO9004:2009

The current issue of the guidance has the following title: *Managing for the Sustained Success of an Organization: A Quality Management Approach.*

Once again, the current version of the guidelines is not intended to be a "how-to" for implementers. A quick review of the topics shows that it's not aligned with the key sections of ISO9001:2008, or with the format and headings of the proposed revision (see the next chapter).

It's perhaps interesting to note that the individual guideline headings include mentions of strategy and policy formulation, deployment, and communication, as well as financial resources, knowledge, natural resources, innovation, and learning. Those familiar with the USA's Baldrige National Quality Award or the European Foundation for Quality Management (EFQM) Excellence Award criteria will notice some distinct similarities.

With the planning and preparation for the next revision of ISO9001 in the very early stages, it is unclear what direction any amendments to ISO9004 will take, or whether any will be necessary. If the usual revision cycle is indeed undertaken, the next version of ISO9004 won't be due for publication until some while after the release of ISO9001, which is slated for potential publication in 2015.

CHAPTER 4: THE FUTURE AND ISO9000

In an earlier chapter, describing the development of ISO9001, it was stated that all ISO documents go through periodic review and possible revision. The cycle has, historically, taken some six to eight years to complete (1987–94, 2000, 2008), and it is likely that the next version will be released sometime in 2015.

The ISO Committee process of identifying revisions, drafting, review, and approvals is defined by the following phases:

- vote to revise;
- new work item proposal;
- plenary meeting – start work;
- first committee draft issued;
- draft international standard (DIS) for ballot;
- final draft international standard (FDIS) for ballot;
- standard issued.

At the time of writing this book, the next review cycle has started and feedback information was gathered by the international organization from users through online surveys etc.

One thing about ISO9001 that seems to be fairly certain to change is the high-level format or layout of the contents.

In order to permit easier integration of ISO9001 with a number of other management system standards – ISO14001, ISO50001, ISO13485, ISO27001, etc. – a singular format has been adopted.

The proposed format should look like this:

ISO9001:2008 current quality management structure	Common management standards structure
1.0 Scope	1.0 Scope
2.0 Normative reference	2.0 Normative reference
3.0 Terms and definitions	3.0 Terms and definitions
4.0 Quality management system	4.0 Context of the organization
5.0 Management responsibility	5.0 Leadership
6.0 Resource management	6.0 Planning
7.0 Product realization	7.0 Support
8.0 Measurement, analysis and improvement	8.0 Operation
	9.0 Performance evaluation
	10.0 Improvement

An area of potential content change is likely to include the replacement of "Preventive action" with "Risk management." Risk management is a feature of other quality management system standards, including ISO13485 for manufacturers of medical devices and AS9100 for suppliers of aerospace equipment.

In 2009, the International Organization for Standardization published a set of standards related to the management of risk, ISO31000. It comprises principles and general guidance on risk management. Unlike ISO9001, ISO31000 is NOT to be used for the purposes of third-party certification as a risk management system. It does, however, outline how an organization may take steps to identify, prioritize, assign authority for, and mitigate risks.

CHAPTER 5: IMPLEMENTATION – A HOW-TO GUIDE

Once an organization has committed to implementing a quality management system, based on ISO9001, often the question arises, How? From what's been described in earlier chapters, it should be clear that implementation of a quality management system in compliance with the requirements of ISO9001 isn't simply "Say what you do, do what you say," or documenting a set of manuals, procedures, and work instructions to suit the various requirements of the International Standard.

Commonly asked questions voiced by those who come new to ISO9001 implementation include, "How long does it take?" "What does it cost?" And "We've been in business for a long time, what else have we got to do to meet ISO?"

If we take a look at a tried and trusted methodology and describe the typical activities as well as key milestones involved to the point where the organization is ready to become certified, it will help any organization to answer – or at least to determine an estimate – what their particular implementation needs would be. If we take an analogy from the new product development process – the processes in which an organization launches a new product into the marketplace – we find that there are some distinct similarities.

The phases and key action items and milestones can be represented diagrammatically, as shown below.

- Needs assessment (also known as a gap analysis)
- Planning and preparation

- System design and documentation
- System implementation and audit
- System review and improvement

(The Certification Process is covered in the following chapter.)

PHASE 5 – SYSTEM REVIEW AND IMPROVEMENT
Management review
Corrective action
Improvement
Customer satisfaction
Data analysis

PHASE 4 – SYSTEM IMPLEMENTATION AND AUDIT
Process implementation
Product and process measurement and monitoring
Internal audit

PHASE 3 – SYSTEM DESIGN AND DOCUMENTATION
Quality manual
Process sequence and interaction
Documented procedures
Other documents necessary
Records identified

PHASE 2 - PLANNING AND PREPARATION
Competency/training requirements defined
Management representative identified
Detailed action list/timeline for implementation
Responsibilities and authorities defined
Quality policy, objective, and measurements identified
Communications plan created
Budget and resource plan

PHASE 1 - GAP ANALYSIS
Three types of gap identified
Management commitment
ISO debriefing/overview session
Draft quality policy, quality objectives

Phase 1 – the gap analysis

Those organizations which make the decision to implement a quality management system in compliance with ISO9001 often have many of the requirements of the International Standard in place – in some way, shape, or form, simply through the needs of doing business and the innate ability of people to create processes, controls, and documentation, and for many reasons.

This fact, plus the commonly held belief that implementing ISO9001 means "Say (document) what you do, do what you say (document)," can lead organizations down the wrong path, leading to too much documentation being created, often in a format which is a difficult to use and maintain (*see Chapter 2*).

Similarly, a "desk analysis" may be performed, comparing what the organization has in terms of "paperwork" – only – against the various ISO9001 requirements. This can also give an incorrect view of the "gap" between what's required by the Standard and to the organization's current situation, since it doesn't take into consideration actual practice and is, therefore, only a two-dimensional view of the quality management system.

A fully effective gap analysis takes a look at the currently implemented practices of an organization in terms of:

- things ISO9001 requires which the organization practices, but in a way which may not be formally or fully compliant with the requirement;
- things ISO9001 requires which the organization isn't practicing; and

- things ISO9001 requires where, whether the organization has formally defined a process or not, implementation isn't fully effective.

To determine the type and extent of these "gaps," an audit may be undertaken by someone fully competent in understanding ISO9001 and effective auditing techniques. This second aspect of competence is of vital importance, since the auditor will be evaluating undocumented processes etc.

Frequently, organizations choose to contract a qualified consultant to perform the gap analysis, since these often possess the necessary skills and experience to carry out the audit, provide the report, and also debrief management in the nature and extent of the gaps as a prelude to the next phase, which involves planning for implementation. It is not uncommon for management to have a (brief) overview of ISO9000, including the background of its development, ISO9001 requirements, and the certification process, given by the consultant after the gap analysis audit has been performed.

Throughout the management overview session, the findings from the gap analysis can be used as discussion points to better describe and explain what's required by the International Standard and, if appropriate, to begin the creation of a detailed action plan. Some form of visual metaphor or graphic can be used to assist in effectively conveying the state of compliance with ISO9001, giving the organization's management a clear picture. For example:

= Requirement not met, no process established or implemented, or improvement of effectiveness required.

= Requirement nearly met, formal definition or documentation required to fully comply.

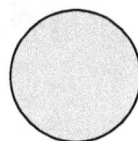

= Requirement met, no further action required.

By way of a caution, it's been known for the services of an ISO9000 certification body to be engaged to perform this gap analysis, for many reasons. Their auditors are, after all, familiar with the requirements of ISO9001. Experience shows that this is not the most effective path to take, since the rules of accreditation restrict the certification body from consulting – including the delivery of on-site or in-house training at a client organization. If the audit is performed by someone other than the person who leads the management overview training event, the contents of the gap analysis report may be difficult to interpret. This, in turn, may result in discussions not being based on the fullest understanding of actual practices.

Furthermore, since a certification body cannot provide consultation services, the organization may not get the fullest benefit of their understanding of the ISO9001 requirements, in the context of their processes etc., during the debriefing. Being told where there are non-compliances with the Standard's requirements is only part of the information the organization's personnel need to know. They also need to know how to effectively close the gaps and what options are open to them in terms of tools, techniques, and other resources.

While undertaking an ISO9001 management overview course, it would be appropriate to begin developing a clear action plan of activities, timing, and responsibilities to close the gaps.

By whatever means this initial phase is accomplished, there should be a number of key deliverables:

• The commitment of top management to implementation of the Organization's quality management system – including commitment to budget, timing, and personnel resources to support implementation.

• A communications plan – what this means to the organization, the roles people will be playing, the timeframe, descriptions of what's going to be new/different, and how it affects people – for example, internal audits, the purpose of ISO9001 certification, and the process (if electing to become certified).

Phase 2 – planning and preparation

The successful implementation of a quality management system will require a detailed project plan of tasks, deliverables, responsibilities, and duration/timing. For each of the subsequent phases the organization must ensure that all "gaps" previously identified are broken down into assignable activities with clear results. By constructing a clear plan, there is a greater likelihood of management being able to set clear expectations, to monitor progress and to address roadblocks. Furthermore, a better understanding of the duration and total resources required can be identified which may translate into accurate budgeting and associated expenditure controls.

A simple project plan created as an Excel® spreadsheet or similar can be an effective tool. *See Appendix 2* for an example.

Whatever format is chosen, the plan for the creation and implementation of the quality management system should clearly define:

- the person(s) responsible for accomplishing the task(s);
- the duration of the task(s);
- the deliverable(s) associated with the task(s);
- the timing of the task(s) relative to others; and
- a clear indication of the task(s) that are related (feed) or result from others, in a logical sequence.

As each task is completed (or not), progress can be tracked, using "traffic light" (R/Y/G) status indicators. Those which fall into "yellow" (behind schedule) or "red" (stalled) status can be brought to the attention of management for understanding and resolution.

Frequent reviews should be undertaken by the personnel involved in the implementation to ensure the plan is kept updated and that any unforeseen activities are added, with the appropriate assignments, etc. As progress is made, the reviews will morph as each phase is undertaken.

Furthermore, it is possible to eventually transition the implementation project reviews into a platform for the management reviews required by ISO9001.

On completion of the plan, approval should be gained from top management to proceed.

Phase 3 – system design and documentation

This phase is very similar to the design and development phase for a product. The "design" of the quality management system documentation is, in many ways, like a product. The quality management system documentation must be planned and have competent people to create it, and consideration must be given to how requirements which are applicable are going to be incorporated. Once designed, the documentation must be produced and assembled.

"Good" product design is generally also accepted to be something which is handed over to the people who are tasked with making it in a seamless manner. The design outputs are understood, well defined, meet any applicable regulatory requirements, and, most importantly, can be implemented – or made – without causing problems. It's therefore important to involve the people who make the product in ensuring that requirements are defined in a way which makes life easy for them, by soliciting their input.

Rarely, and unfortunately, this comparison is never fully made, with the result that documentation causes problems for the people who have to use it. It's been stated, previously, that more attention is given to "tiers" or levels of documentation, or to the format, and to laying out things by ISO clause than to the veracity of the information contained in the document.

It is of vital importance that the people who do the work contribute to the creation of whatever instructions are really needed to perform work. It is often the case that having created a document describing how some work is to be carried out, it is "thrown over the wall" to the supervisor to ensure training is carried out.

In creating documentation for the quality management system, a good starting point is to inventory each and every existing document. Because these are documents which people find somewhat useful – even if not "perfect" – they can be used when completing another important step.

Mapping the processes of the organization is a very revealing part of the creation of the quality management system and its documentation. Process mapping should be performed by those members of staff who are involved with each individual process, including the process owner, the internal "supplier," and the "customer," facilitated by an impartial person. This impartial person is primarily there to ensure that no one person takes a lead in describing the process, that the existing process is captured "warts and all." Capturing the process "as is" is significant because opportunities to improve what happens can be readily identified, particularly where waste occurs.

It is tempting for an individual to sit down and "flow-chart" a process, using the type of flow diagram used in determining the logic behind early computer programming. Apart from anything else, such a technique captures only what the "logic" is, not what actually happens.

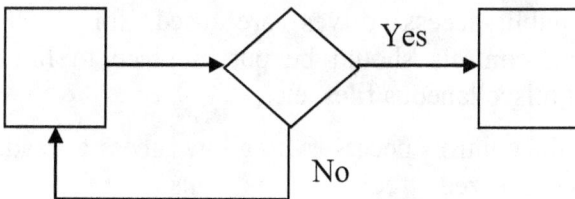

As discussed previously regarding the format of documentation, effort may be expended on the shape of the elements of the flow chart and their relative meaning, rather than on gathering information on the actual process, possibly to the overall detriment of the result.

Capturing each process in its rough form can best be achieved by papering a wall of a meeting room, free-form, using a variety of markers, "Post-it®" notes, plain postcards, etc.

Using examples of the existing documents which have been inventoried, the process maps can be "populated." As issues are detected, consideration may be given to the contribution of any documents. For example, if a purchasing requisition form is used, but it's not clear how much information is needed (as a minimum), then the buyer who receives it may have to halt the process, causing delays, etc. This should become apparent and a change to the form design or a clear instruction on what the minimum criteria are to be should be given.

Once captured in "rough" form, consideration can be given to the final format for documents. With networked computers and a variety of applications such as Microsoft® Word®, Adobe Acrobat Reader®, and Microsoft® Visio®, the creation and structuring of the quality management system documentation can be logically defined. Care should be taken, however, to ensure a convention for the naming of the various files and folders. Experience shows that when public-access drives are used for QMS documentation, controls should be put in place to limit "dumping" of miscellaneous files, etc.

Furthermore, the controls necessary to allow access to read, without unauthorized editing, is also relatively straightforward to accomplish. One consideration is that unless all personnel who participate in the quality management system have access to a computer or tablet to view the documentation, some form of hard copy documentation may still be required by those people.

At some point in the development of the documented quality management system, it will be important to "put down the pens" and begin the fourth phase.

Phase 4 – system implementation and audit

For a proportion of the processes, it will be "business as usual," since the creation of the quality management system has merely formalized what was already in place and working well.

For those other aspects of the quality management system which are new or even substantially changed – reference back to the gap analysis results – it may be necessary to identify new or additional competences. As has been stated previously, it isn't absolutely necessary to train people on everything new or different. What should be established is whether anyone involved in performing a process – or part of it – is required to possess any additional or different competences.

An example of a "knee-jerk" conclusion being drawn – which results in training – is when identifying candidates to become internal quality system auditors. While it may be obvious to suggest that most organizations have no one who possesses the necessary skills and experience as an internal auditor, sending a handful of people to a two-, three-, or even five-day auditor training course may not result in all the appropriate competences being developed.

At best, such training permits only a very basic level of audit skills.

Consideration must be given to some other basic or "entry-level" criteria for auditor candidates, onto which the auditor skills may be built. Even fundamental communications skills are often overlooked, as are interpersonal or "soft" skills, and even knowledge of the purpose and operation of other functions of the organization. The International Standard on Auditing, ISO19011, gives some excellent descriptions of characteristics auditors should possess.

When the processes of the quality management system are demonstrating results, it will be necessary to perform internal audits.

As has been established, there are many and varied myths surrounding establishing and implementing an internal audit program. When faced with determining what and when to audit the organization's newly minted quality management system, there's a clue in the requirements from the ISO9001 Standard's clause 8.2.2. This states, in part, "that an audit program (a number of audits) be planned, taking into consideration the status and importance of the process and areas to be audited." Substituting "risk and importance" for "status and importance," we can develop an understanding of where and what to audit as a priority. In the context of the quality management system, we can take risk as being higher with new or changed aspects and importance as being that which has an impact on customers, regulatory compliance, or – in the early stages leading up to certification – compliance with ISO9001. It would be entirely appropriate to select those processes/areas/activities of the quality management system which are new to the organization, have been changed, and have a potential impact on customer satisfaction.

Phase 5 – system review and improvement

Following the period of implementation and auditing of the quality management system, a point is reached when there is sufficient data regarding the performance of the processes. The next key milestone is to prepare and perform a review by the organization's management.

From the performance data of the various processes, it will be necessary to analyze this for trends, etc. This should be performed by the various process owners, including identification of corrective and improvement actions.

Despite ISO9001 not requiring a meeting in order to perform a review by management of the quality management system, experience has shown that this method brings a number of benefits. In many situations, the results of the internal audit process are not well understood as a review item. Since the internal audit process, for example, is new to the organization, it is common for the review to go through a detailed analysis of the findings and so on without looking at what they really mean.

By looking at the following matrix, we can see that internal audit results can be viewed as validation of the QMS being effective in producing process results. Although a clear set of review topics is defined in ISO9001, what is often missing is a "30,000-feet view" of the QMS as a management tool to deliver results.

Process performance	Internal audit results	Actions required
Meets or exceeds targets set	QMS being followed	Improvement of QMS?
Meets or exceeds targets set	QMS NOT being followed	Improve QMS in line with practice
Below targets set	QMS being followed	Corrective action required on process
Below targets set	QMS NOT being followed	Corrective action on QMS

Each process owner brings his or her analysis of their process performance, objectives, and action items to present and discuss with the other process owners.

From the review, opportunities to improve some aspect of the quality management system should be weighed, the appropriate resources identified, and assignments made to roll out the approved actions.

One of the outcomes of holding a review of the quality management system is to confirm the readiness to undertake the certification audit. When all process owners can confirm that they know their process performance, and have taken actions to correct and improve some aspect of the quality management system, so that (importantly) the internal audits have validated the use of the quality management system in controlling the processes in order to achieve results, the organization is substantially ready to undergo the Stage 1 certification audit.

Once certified, the review of the quality management system will be substantially similar. However, part of the review will take into consideration preparation for, and results of, subsequent certification body audits.

Once an organization achieves certification of its quality management system, the focus should be turned from the basic implementation needs and toward maintenance and improvement. By referring back to the "Plan, Do, Check, Act" cycle diagram, we can see that the organization can employ key activities of the quality management system in directing the maintenance and improvements required.

As stated previously, management review of the quality management system is the cornerstone of maintenance and improvement.

Once certified, the review ... ... finalize that ...
system will be substantially ... ... the
review will take into consideration ... ... other ... ... and
results of public ... certified ...

... for the ... ... quality
management system ... ... setting
basel... information ... ... finance and
... ... By ... ... in ... Part Two
... cycle ... ... thereafter ...
... ... key activities of the ... ... will ... explore
... ... the ... ... and on ...

... ... provide ... manage ... ... of the ...
... ... the ... ...
in ... ...

116

CHAPTER 6: THE CERTIFICATION OF MANAGEMENT SYSTEMS

This chapter describes the background to the development of so-called "third-party certification of management systems," or what's more commonly known as "ISO certification." This term is, in fact, somewhat misleading since the ISO organization doesn't involve itself in any aspect of the certification process.

Although not required to be successful in implementing an ISO9001-based quality management system, an independent certification of compliance is a common option for many organizations. Often, major purchasing organizations in the aerospace, automotive, defense, and medical device industries require their direct suppliers to obtain an independent certification to ISO9001 as a minimum, before contracts are awarded. This requirement is often handed down the line to lower-tier suppliers.

Unless specifically required to comply with customer or regulatory requirements, an organization may either choose a self-declaration of compliance with ISO9001 or choose to have a significant customer attest to the same. These aren't usually as acceptable to discerning customers, or for other reasons, as the option of an independent certification audit performed by an (IAF-accredited) certification body/registrar.

Before we scrutinize the role and processes of a certification body, it is worth spending some time understanding aspects of accreditation or oversight which govern their operation in the marketplace.

The importance of accreditation of certification bodies

Accreditation of certification bodies is seemingly not well understood by those who have to employ certification services, despite it being of vital importance to the credibility and quality of the resulting audits. Some basic information about the influence that accreditation has on the certification body and its processes is very important to any organization considering offering their QMS for certification. If accreditation of an organization's ISO9000 certification is not recognized by a customer, it may mean starting back at square 1, with the costs being doubled.

Accreditation of certification bodies started in the UK in the late 1980s, shortly after ISO9001 was made publicly available. In response to a (1977) UK Government white paper, written by Sir Frederick Warner, entitled "Standards and Specifications in the Engineering Industries," which identified the opportunity to reduce the costs and disruption associated with multiple supplier audits of the same suppliers, the concept of using an independent body to certify compliance with the (appropriate) international standard was floated.

To ensure a minimum level of consistency between what became known as "certification bodies," the UK government established an oversight or accreditation body titled the National Accreditation Council for Certification Bodies (NACCB). A similar accreditation body was also established in the Netherlands, known as the Raad Voor Certificatie (RVC). These organizations are currently known as UKAS (United Kingdom Accreditation Service) and RVA (Raad Voor Accreditatie), having changed their names to better describe their activities. Other nations followed suit as ISO9000 gained acceptance around the globe, including the USA, where today the American national accreditation body is run by the American National Standards Institute (ANSI).

Accreditation bodies evaluate and monitor the performance of certification bodies (as well as other, similar organizations) against a set of defined requirements, from another ISO standard, this time ISO/IEC 17021. They ensure that the audits are carried out of defined processes and criteria, including the competence of auditors. Applying ISO/IEC 17021 to a certification body is not unlike an organization implementing ISO9000 for their products. Both standards are employed to ensure a quality output from the organization's business processes.

Today, accreditation of certification bodies is managed on a national basis, by an organization representing each country. To ensure consistency across nations, accreditation bodies may subscribe to the International Accreditation Forum (IAF), which, through multilateral agreements, provides oversight of the certification bodies. In addition to the requirement of ISO/IEC 17021, the IAF also publishes a number of documents, including tables which define the number of days an audit should take, based on the number of people (headcount) involved in the organization's QMS. For example, the IAF's MD5 table can be found on the IAF.org website and also includes descriptions of the factors which allow for tailoring of audit time. Organizations considering certification can obtain "insider information" to ensure that their audit duration is appropriate.

Certification body services

When selecting a certification body (CB), the default supplier selection process often starts with getting three quotes. Quotes may be solicited from these three certification bodies, and, once they're received, a review might reveal ... nothing, except that the amounts of time for the audit (in days, usually), and therefore costs, appear to be very similar. Since the IAF MD5 document defines the duration of each audit, based upon the headcount of the organization, the only other costs associated with the auditing service will relate to the following (typical) factors:

- auditor daily rate;
- fees, labeled "admin" or "account management"; and
- fees for reviewing corrective actions arising from any nonconformity reports issuing from an audit.

Possibly, the most significant of these costs is the auditor daily rate, since this may be a key indicator of the amount the auditor charges the CB (many auditors are subcontractors, not employees), or perhaps how much the CB is prepared to pay the auditor. Many auditors are listed by the body that they are certified through, e.g. RABQSA, in an online directory. It may be that the chosen CB selects a candidate auditor only from that list, without consideration of any other characteristics than the qualifications needed to be on the list!

Another question to be asked might also be whether the auditor is local to the client's premises. Although an auditor's being "around the corner" is optimal in keeping travel expenses lower, this doesn't consider the following two facets:

- The auditor who is local may not possess the necessary experience (EAC code) to suit the client's business type.
- There may be an auditor who is scheduled in the area to perform other audits and, therefore, travel expenses can be amortized across a number of clients. If a client is flexible about audit dates, such an arrangement can work well.

It's not unusual for an organization to send a list of questions to be answered by certification bodies, as input to the selection process. Below are some typical points which have been asked of certification body candidates:

- Who you are accredited with, and are they accredited by a signatory to the IAF, or ANAB-accredited?
- Do you publish an official interpretation of the ISO9001 requirements and can we get a copy?
- Have you ever had your accreditation revoked or suspended? If so, what were the circumstances?
- Share the number of local clients in our area.
- Detail your experience in auditing facilities like machining job shops.
- Share the number of local lead auditors in our area.
- Detail the experiences of these lead auditors in auditing facilities like machining job shops.
- Detail the number of lead auditors available for the auditing and maintenance of our certification.
- How flexible are you on scheduling or rescheduling?
- How do you handle differences of opinion over interpretation of the Standard?
- What is the process of handling differences, and whom do we contact?
- Is the stage 1 audit conducted on-site, off-site, or a combination?

- How fast can we set up an interview with you?
- Can we also set up an interview with the lead auditor that would be assigned to us?
- Does your accreditation cover our organization's industry?

If we take a look at these questions, many could have been answered at the same time as question one was answered – and that could be simply answered by a quick check of the IAF website! Accreditation to ISO/IEC 17021 takes care of issues such as auditor qualifications, appeals, conduct of the stage 1 audit, and so on.

Perhaps it's interesting to note that such questionnaires completely miss many of the points which really affect the actual relationship an organization will likely have with its chosen certification body! Rather than waste time asking questions which are actually answered by their accreditation – which is a "playing field leveler" – it would be better to address the actual performance of the certification body in delivering more than just an audit and certificate of compliance with ISO9001!

Let's consider what aspects of the services provided by a CB impact the organization:

- quality of service (technical, scheduling, value of audit reports);
- quality of auditors (competence, industry experience, audit approach, professionalism);
- credibility or reputation of certification (how do your customers and other certified clients perceive the CB?)
- what's the "Scope of Accreditation" and does it cover the organization's type of business?

Having selected a certification body, the process begins with providing information about the organization on which a quotation may be based.

Typically, the information required by the CB, on which they base their quote, includes:

- Contact personnel details.
- Names, positions, phone numbers, email addresses, fax numbers.
- Organization details:
  o Name, street address (HQ).
  o Other locations (if applicable).
- Head count involved in QMS.
- The scope of the QMS.
- Exclusions from the QMS, such as design, customer property, control of measuring equipment.

The quotation or proposal is likely to reflect the amount of audit time for the following activities:

- the "stage 1" audit,
- the "stage 2" audit,
- the surveillance audit (usually annually), and
- the "triennial reassessment."

It is common for an ISO9001 certificate to be valid for three years. This is a legacy of the approvals given by the UK's Ministry of Defence (MoD) to their suppliers. Early adopters of third-party certification, in the early 1990s, were often suppliers to the UK MoD, the British government, and various government departments, making it a requirement for them to be third-party certified.

Note – if a CB doesn't quote a cost and duration for the triennial audit, the guidance from ISO/IEC 17021 states that the audit is approximately two-thirds of the combined stage 1 two audit durations.

The certification body will also issue some form of contractually binding agreement. This describes the terms and conditions relating to payment for their services, and details of the organization's commitment to the rules relating to certification of their quality management system. These rules are passed down, from the accreditation bodies, by the certification bodies, to their clients, and are defined (in part) in ISO/IEC 17021. Typically, the rules include:

- confidentiality of the information obtained about the organization;
- changes relating to the organization's
  o ownership,
  o management, and
  o locations;
- changes relating to the
  o scope of the QMS in terms of products, locations for example, and
  o major changes to the QMS, including those affecting management, regulations, documents and so on; withdrawal of certification.

An additional audit service offered by most certification bodies is known as the "preliminary assessment" or "pre-assessment." This is a purely optional audit which is not part of the formal certification audit process; is usually a shorter-duration audit; and, as its name suggests, is performed before the actual certification audit is performed. This audit is described later.

Certification body audit process – the basics

Before the various types of audit conducted by a certification body are described in detail, it's worth taking a look at the basic activities at the core of these audits. These activities are normally founded on the requirement of ISO19011, "Guidelines for Auditing Management Systems."

The key components of the audit process include the following:

Opening meetings

At the scheduled time, according to the audit schedule/agenda (or other arrangements made), the assigned lead auditor will chair a meeting with representatives of the organization. An agenda for the meeting might look something like this:

- Introductions of the auditor(s) and the organization's participants.
- Purpose of meeting and audit scope.
- Review of audit agenda/changes.
- Logistics: work space, meal arrangements, working hours, etc.
- Audit guides.
- Audit process: interviews, evidence gathering, etc.
- Audit reporting: verbal, written, grading.
- Daily reviews (if multi-day audit).
- Confidentiality statement.
- Closing meeting purpose and timing.
- Discussion/Q & A.

Audit activities

On completion of the opening meeting, the auditor(s) will begin their audit assignments. The audit purpose is to verify that the organization's quality management system is implemented and is effective in achieving the stated objectives. They must "test" various peoples' understanding of their jobs and process controls, and determine that the organization is effectively planning for the results they and their customers expect to achieve.

To accomplish this, auditors will interview the relevant personnel, from management to members of staff and associates who perform work, asking questions about the processes they work on, the objective(s) of those processes, and current performance of those processes. The auditors will likely also ask questions which are intended to verify understanding of various peoples' responsibilities and authority for control of processes, including taking action when unplanned situations arise, perhaps resulting in nonconforming products, etc.

Auditors make copious notes as they verify the evidence they see and hear, including references to the specific documents and records they request to see. These records will, typically, include customer orders/contracts, purchase orders placed on suppliers, competence evaluations and training records, minutes from product design review technical meetings, internal auditors' notes, and so on. These are compared to the organization's documented quality management system and, with what was learned from interviews, a picture is formed by the certification body auditor(s) as to the degree of compliance and effectiveness.

Closing meetings

At the conclusion of a certification body audit, whether it's stage 2, surveillance, or triennial reassessment, the auditor will convene a "closing meeting." The purpose of the meeting is to summarize the findings of the audit for the organization, to discuss any follow-up actions which may be necessary, and to outline the purpose and timing of the next visit.

At the end of the audit, one important action for the auditor is to deliver a recommendation to the organization, appropriate to the type of audit which has just been performed. The recommendation is based upon the evidence gathered and conclusions arrived at by the auditor(s).

Example recommendations are defined in the following descriptions of each audit type.

Audit reporting

Certification body auditors are required to fully report the results of the audits they perform. There are two basic forms of report:

- nonconformity reports,
- audit summary reports.

Nonconformity reports (NCRs)

Myth alert!

Three or more minor NCRs make a major NCR.

A major NCR means the audit is over.

Nonconformity (nonconformance) reports are possibly the most interesting to the organization being audited. As the audit unfolds, the auditor may observe a situation or situations where evidence indicates the quality management system is not being implemented, or is not as effective as intended or planned.

In these situations, having agreed the facts with the organization's representatives, the auditor will complete a nonconformity report (form) with the following fundamental information:

- the source of the audit requirement,
- the audit requirement,
- the source of the audit evidence, and
- the audit evidence observed.

An example nonconformity report statement might read as follows:

The organization's Quality Manual, revision #3, states in paragraph 3.3.1, that management reviews are held with a minimum of three Vice Presidents (Engineering, Production, and Quality) in attendance at a meeting to discuss process performance to objectives.

The minutes (record) of the review meeting held on January 31, 2012, indicates that only two VPs were in attendance (Engineering and Quality). As a result, there was no review of Production-related performance to objectives.

The nature of the content of each nonconformity report is reviewed and graded by the auditor. Grading gives "gravity" to the content of the report as a means to communicate significance. It is typical that a certification body auditor will consider whether what's been observed is an isolated or localized nonconformity, not indicating a systemic issue, or whether the situation is determined to be a breakdown of the quality management system. Such a breakdown might be a failure of effectiveness, a failure to implement, or a significant number of the (initially) localized nonconformities clustered around a specific ISO9001 requirement.

Myth busted!

In an ISO9001 certification audit, there are no specific "rules" for determining when a number of observed nonconformities warrant classification as either "minor" or "major." A certification body may have its own definitions of categories of nonconformity reports, including "major," "minor," "opportunity for improvement," "category 1," "category 2" and so on, and some even have definitions of how many "minor" nonconformities found constitute a "major" nonconformity, in their certification service agreement or contract.

Audit summary reports

The certification body auditor is required to demonstrate that a comprehensive audit has been carried out, so, toward the end of the audit, a summary of the audit is completed. The certification body will provide the auditor with a form in a prepared format which is then filled out with the details of the specific audit. Details are completed based on the notes taken, people (job titles) interviewed, records reviewed, and so forth.

The summary report will also contain a recommendation to the certification body's management on the status of the organization's QMS and whether a certificate of compliance should be issued.

The preliminary or pre-assessment audit

Since most organizations come to ISO9001 certification without significant experience of the audit process, the preliminary assessment can be a very useful experience. For readers who are familiar with the performing arts, it's quite normal, immediately before the "first night" of the performance, for the performers to have a "dry run" through their performance, in the venue, with other performers, musicians, etc. This dry run is often referred to as the "dress rehearsal" and is done to make any final adjustments to the performance before the public get to see and critique it.

A dress rehearsal isn't to make major adjustments to the score, choreography, costumes, etc., since there's no time available. Instead, it gives the producers an opportunity to visualize the performance *in situ*, and, hence, minor adjustments may be identified and accommodated.

Unlike the "stage 1" and "stage 2" audits, there is no defined duration for the preliminary assessment, so the organization can choose how long it believes is needed. What's more, the organization gets to decide what they want the auditor to review during the visit. The agenda is theirs to define. An organization may decide to have the auditor focus on a few aspects, or take a sweep of the entire quality management system. In selecting a smaller focus, organizations often are interested in those aspects of the QMS which may be new to them, e.g. a calibration system, where none was formally defined or implemented before.

Often, a broad view of the status of the system as a whole can validate for the management team that their efforts are, indeed, ready to undergo the more detailed scrutiny of the registration audit. It acts, therefore, as a dress rehearsal for the actual certification audit.

Other benefits include:

- Observing how the assigned certification body auditor goes about performing the audit.

Although the organization will have conducted internal audits, it's always good to know how the CB auditor does things.

- Observing how they interact with the various people they interview.

Each CB auditor is different and they have unique ways of establishing good communications with the people they interact with. It helps to "break the ice" if you know this ahead of time.

- Allowing various (key) people of the organization to experience being audited by the certification body auditor.

As before, although internal audits will have been performed, not everyone will have had a role in those. Some people might be natural candidates to be audited by the CB auditor, and it's a great time to give them that experience.

- Uncovering a potential weakness in the system before the stage 2 audit.

This is very useful, of course, as the auditor who did the preliminary assessment will be very aware of the actions you took to rectify the situation found, which also builds confidence in the commitment to implementing an effective system.

The preliminary assessment can be timed with the stage 1 audit. This allows for some continuity of understanding for the auditor since they will have been able to study your system documentation and, while that knowledge is still "fresh," to take a look at aspects of the implementation too. It also helps to cut down on expenses!

The results of the preliminary assessment can be a useful input to the "management review" as a formal indicator of the "suitability and effectiveness" of the quality management system, in the days leading up to the registration audit. Any report from the preliminary assessment is not supposed to have any impact on the actual certification audit, it being entirely up to the organization's management to determine if any of the auditor's reported comments are significant enough to warrant corrective actions. In actual fact, if the same auditor who did the preliminary assessment performs the stage 2, then, if the client chooses to act on the auditor's comments and observations, it will be noticed and will possibly be viewed positively by the auditor, as validation.

A recommendation from the preliminary assessment is whether the client's QMS is in a suitable status to be successful at the certification audit.

The stage 1 audit

On a mutually agreed date, the certification body's assigned auditor (also known as the "lead auditor") performs what is known as the stage 1 audit. The purpose of carrying out this audit is defined in ISO/IEC 17021 as being:

- To evaluate the documented quality management system.
- To ensure that the Standard's requirements have been understood and that key performance, processes, objectives, and operation have all been identified.
- To verify the information collected regarding the quality management system's scope and processes, the locations of the business, and any related regulatory and statutory compliance issues.
- To review and agree the duration and timing of the stage 2 audit.
- To plan the audit activities of the stage 2 audit.
- To ensure that the internal audits and management reviews have been carried out and that there's sufficient implementation to support readiness to undergo the stage 2 audit.

The stage 1 audit includes a report detailing the above, plus any findings from the audit of documentation etc.

As well as this report, another key output of the stage 1 audit is a plan or agenda for the stage 2 audit. That plan will, typically, detail each individual focus of the quality management system with the duration, timing, and auditor assignments. It may look something like this example (for an organization designing and manufacturing pumps) for a team of two auditors:

Monday, April 4	Lead auditor – audit activity	Auditor 2 – audit activity
8.00am	Opening meeting with management	
8.30am	Overview of QMS, objectives and key measurements	
9.00am	Proposals and contracts processing*	Product design and development*
10.00am	Production planning	
11.00am	Purchasing process, incl. outsourcing controls*	
12.00pm	Lunch	
12.30pm	Receiving and inspection*	Quality planning and controls*
1.00pm	Manufacturing process – housing machining*	Manufacturing process – casting*
2.00pm	Pump assembly*	Manufacturing process – impeller machining*
2.45pm	Pump testing*	Inventory control and warehousing*
3.15pm	Nonconforming product controls*	Pump finishing*

4.00pm	Review and compilation of day's findings	
4.30pm	Presentation of day's findings and day 2 schedule/adjustments	
5.00pm	Auditors depart	
Tuesday, April 5	**Lead auditor – audit activity**	**Auditor 2 – audit activity**
8.00am	Calibration of measuring equipment*	Pump packaging/ shipping
8.45am	Customer feedback and satisfaction*	Equipment maintenance*
9.15am	Internal audits*	Personnel training programs
10.00am	Corrective action*	Product and process improvement*
11.00am	Continuous improvement activities	
12.00pm	Lunch and discussions	
12.45pm	Management review	
1.00pm	Audit findings analysis and report preparation	
2.00pm	Closing meeting	
3.00pm	Auditors depart	

* Audit activities will also consider relevant controls of documentation, records, product status, measurements, monitoring, personnel competences, and data analysis.

Once the stage 1 audit is completed, the organization should be clear about the next steps and timing of the stage 2 audit, plus any actions necessary to address issues arising from the stage 1 audit. Typically, this means taking corrective actions on items raised by the auditor which can affect the successful outcome of the stage 2 audit. A formal submission of actions to the certification body is not usually required, since they will be verified as part of stage 2.

The recommendation from this audit relates to the state of preparation of the client's QMS to undergo the stage 2 audit, within the agreed timeframe.

The stage 2 audit

Myth alert!

Certification cannot be granted unless nonconformity reports are closed.

If the auditor(s) find a "major" nonconformity, the audit is over.

Dates for the stage 2 audit were probably agreed and arranged by the auditor with the Organization's management during the stage 1 audit. When the day arrives, the auditor – possibly with a team of auditors (dependent on the audit duration, etc.) – arrives to perform the actual audit of the implementation of the organization's quality management system.

Commencing with an opening meeting (described earlier), the auditor(s) will follow the defined plan, accommodating any unplanned adjustments as needed while still ensuring that all the requirements are covered and that the organization's quality management system is fully assessed.

At the end of each audit day, it is typical for the results of the audit to be presented in a debriefing session. This gives an opportunity for any personnel who were not involved in the audit to hear what has been found, if anything, and to learn about what progress has been made and any deviations to the plan or changes for the next day(s). It is also an opportunity for the auditor(s) to indicate any potential audit trails which need to be followed to verify evidence found or to discuss the potential for any nonconformity to be mitigated based upon better explanation of the implementation evidence discovered.

At the end of the audit, the auditor(s) compiles a comprehensive report for submission to the certification body. The report is summarized at the closing meeting, (defined earlier).

Myth busted!

Although there is the possibility for one or more "major" nonconformities to be discovered during the stage 2 audit, the (lead) auditor will generally advise the organization's management as soon as it is found. Remember, a "major" nonconformity will prevent the auditor(s) from making recommendation for certification to be granted. Although the organization has the option to terminate the audit at this point, it is recommended to continue with the stage 2 audit for these reasons:

- the audit has been paid for, the certification body won't refund any audit time unused, and
- the audit should be completed as planned to reveal all existing issues, so they may be corrected.

Myth busted!

Although nonconformities may be discovered during a stage 2 audit, a recommendation can still be made by the auditor(s) for certification if none are of "major" significance (see previous section on grading of audit nonconformities).

The surveillance audit

On successful certification of an organization's quality management system, the organization moves into the "maintenance phase" of their certification. As defined in the registration agreement, the certification body is required to perform an audit at least annually. If sufficient time is required, a client may elect to have their surveillance audits conducted every six months, to the same total time.

The duration of the surveillance audit is defined in the IAF MD-5 document and is (typically) one-third (33%) of the duration of the stage 1 and stage 2 audits combined.

The surveillance audit is intended to focus on those aspects of the organization's quality management system which are closely related to the maintenance activities:

- any changes made to the quality management system,
- management review(s) held since the previous audit,
- internal audits conducted since the last audit,
- corrective actions,

- improvements,
- customer feedback/complaints, and
- previous audit nonconformities issued by the CB and associated actions.

Over the two surveillance visits, the quality management will be sampled, in particular those aspects which are new and/or changed.

The surveillance audits also have opening and closing meetings and are reported in a similar manner to the certification audit. At the conclusion of the audit, the possible recommendation(s) relate to the ongoing certification or, in the case of significant audit nonconformities (see "major nonconformities"), the recommendation may be to undergo a "special visit" audit.

The triennial reassessment

Myth alert!

After three years of certification to ISO9001, you have to go through a full re-audit.

On the third year after the date of the original certificate of compliance was issued, a "triennial reassessment" is usually performed. Because the organization's quality management system should be somewhat mature at this point, the certification body audit should focus on effectiveness or on corrective and improvement actions which have been taken over the previous years. With some three years of data available from customers' feedback, process performance, product conformity data, supplier evaluations, and so on, the organization's management should have a clear picture of the suitability of their quality management system as a tool to support growth.

As a result of employing a different focus during the triennial reassessment, the duration of this audit is less than the original certification audit. In fact, the MD-5 document allows the duration of the triennial reassessment to be two-thirds (66%) of the stage 1 and stage 2 audit durations combined.

On completion of a successful triennial reassessment, a recommendation may be made to issue a new certificate and the normal pattern of surveillance visits is implemented.

The "special visit" audit

There are circumstances under which an organization's certification body may require a "special visit" audit. The most significant event which leads to such an audit is when a "major" nonconformity is reported during the period of certification. As defined previously, a major nonconformity indicates that a failure in the quality management system has occurred. The nature of this type of systemic failure will usually require a significant corrective action to be undertaken by the organization and, rather than simply relying on records of the results, the certification body auditor may decide to carry out some on-site evaluation of the implementation which led to the records being produced.

Conducted in the manner described earlier, the auditor will focus on the (corrective) actions necessary to address the situation surrounding the major nonconformity. Additionally, the audit will also assess management review and internal audits as key indicators of the manner in which the corrective actions were managed.

At the conclusion of the special visit audit, one of three recommendations may be made and reported to the client, depending on the evidence presented by the client:

- The major nonconformity has been removed.
- Downgrading of the major to a minor nonconformity. Usually, a request will be made for further evidence of implementation of corrective actions to be provided at the next scheduled audit.
- Suspension of the organization's certificate of compliance. If a client hasn't taken (suitable) corrective actions to reduce the status of the major nonconformity, for whatever reason, the auditor is required to initiate suspension of the organization's certificate of compliance.

What qualifications make a certification body auditor?

In the Western world, CB auditors are generally qualified based on a few important criteria, often set by the certification bodies which employ them, including the following:

- RABQSA or IRCA status, which is generally lead auditor or higher status, depending on the scheme categories, and
- experience in defined industries (EAC codes).

Being a certified lead auditor under one or other of these internationally recognized schemes (RABQSA and IRCA) is a professional auditor qualification and is based on meeting a number of criteria, including:

- attendance at an accredited 36- or 40-hour lead auditor course,
- passing the course examination,
- educational and workplace experience in an acceptable discipline, and
- audit experience both as an audit team member and as team leader.

Myth alert!

I've attended an accredited lead auditor course and obtained a certificate, so I must be a "certified lead auditor."

Certificates issued from such a course only indicate attendance and, possibly, a successful examination result, and NOT that the other criteria have been met. For someone to legitimately claim to be a "certified lead auditor," they must have completed the relevant application, including providing evidence of meeting the applicable education, experience, and course criteria; furnished personal references; and, of course, paid the relevant fees!

APPENDIX 1: DCR FORM

Date:	Requester:	D/DCR #

Document type: New ☐   Modified ☐   Deviation ☐   Obsolete ☐

Document ID _____

Reason for deviation/change:

Proposed deviation/change(s): (marked up copy attached ☐)

Reviewed and approved:	
Implemented by:	
Date implemented:	

Document distributed to:	

APPENDIX 2: SAMPLE PROJECT PLAN

For a downloadable PDF version of the following project plan, please visit *www.itgovernancepublishing.co.uk/Publications/Resources.aspx*

Organization's Name
5 Step Action Plan and Responsibilities
ISO 9001:2008 Implementation Project

Organization:	VP	In Progress/On schedule
Mgmt Rep		Overdue - with recovery plan
Current as of		Late - Immediate Recovery Plan Required
# Procedures reqd		
# Procedures Mapped		
Procedures in Doc control		
Procedures implemented		

Steps	Action	Status - Required / Complete	Due Date	VP	Management Rep	Process Owner	Process Owner	Process Owner	Process Owner	Process Owner	Process Owner	Process Owner	Process Owner	Process Owner	Process Owner	Consultant
		Phase 1 - Planning														
Resources	Responsibilities of															
MILESTONE	Management Representative defined & communicated to anyone															
MILESTONE	Confirm role of Management Representative															
	Schedule Project Kick off meeting with staff.															
	Communicate name of Management Representative & duties to staff															
	Assign staff to Steering Committee															
	Establish regular bi-weekly meeting schedule for Steering Committee															
	Letter to all staff stating commitment to ISO 9001 project, registration plan & completion dates															
	Conduct interviews with staff members															x
Budget	Establish budget for consulting, training and related expenses															
	Establish budget tracking															
MILESTONE	Assign Management to roles of 'Process Owners'															
Training	Develop plan for training of people in Quality management system and identify functional training required															
Strategy	Review/establish Process measurements															
Objectives	Establish functional objectives for each area of the business															
		Phase 1 - Planning (continued)														
Quality System	Create Scope statement and justification for exclusions															
	Compile matrix Quality System vs ISO 9001 Requirements															
	Update & get approved the Organizational Chart															
	Maintain up to date job descriptions and/or posting requirements for all positions															
Project Plan	Update this project plan after each Steering Committee meeting															
		1 - Planning (continued)														
MILESTONE	Establish 'War Room' communication for implementation project															
MILESTONE	Gap Assessment of existing system															
MILESTONE	ISO Project 'Go ahead' given, based on plan, timing and															

145

Appendix 2: Sample Project Plan

Phase 2 - Documentation		
Training		
Existing document at ion	I nvent or y exist ing document at ion	
	Compile document at ion against I SO 9001:2008 r equir ement s	
	Dr af t Qualit y Manual	
	Review Qualit y Manual	
MI LESTONE	Appr ove QM	
Deliver able	I ssue QM	
	Dr af t Document Cont r ol Pr ocedur e	
	Dr af t Recor ds Pr ocedur e	
	Cont inuous I mpr ovement	
	Cust omer Sat isf act ion	
	Tr aining	
	Resour ce Planning & Allocat ion	
Pr ocess I mpr ovement	St eer ing Team t o ident if y impr ovement opport unit y/ies	

Phase 3 - I mplement at ion		
Training	Review cur r ent assignment s and assign per sonnel t o f unct ion as Lead and I nt er nal Assessor s	
	Tr ain Assessor s Lead and I nt er nal Assessor s on use of cont r ol sof t war e	
	Ment or Lead and I nt er nal Assessor s	
Assessment	Develop scope f or per f or ming Full Syst em I nt er nal Assessment s	
MI LESTONE	Develop I nt er nal Assessment s Schedule f or t he next 6 mont hs	
	Per f or m depar t ment al I nt er nal Assessment s	
	Per f or m a Full Syst em I nt er nal Assessment s	
	Lead Assessor document s CAR' s in Syst em 9K	
	Per f or m a Full Syst em I nt er nal Assessment	

4 - I nt er nal Audit & Management Review		
	Ensur e all process owners understand Management Review Gov. Proc.	
	Ensur e CAR' s have been addr essed ef f ect ively	
	Ensur e FAME dat abase is updat ed f or CAR' s issued	
MI LESTONE	Schedule Mont hly Management Review Meet ings f or next 6 mont hs	
MI LESTONE	Agr ee Mangement Review Meeting For mat and Agenda	
MI LESTONE	Fir st Management Review Meet ing held.	
	Fir st Management Review Meet ing minut es dist r ibut ed	

5 - I ndependent Qualit y Syst em Assessment		
Planning	Management Repr esent at ive agr ees 3r d. Par t y Pr e & Final Assessment dat es wit h I QA mgr / Regist r ar	
Pr e- assessment	Regist r ar per f or ms pr eassessment	
MI LESTONE	Ver if y t hat all pr e-assessment f indings have been r esolved.	
Syst em Validat ion	Ver if y t hat all int er nal assessment f indings have been ef f ect ively r esolved	
MI LESTONE	Management Team conf ir m r eadiness st at us f or Cert if icat ion	

5 - I ndependent Qualit y Syst em Assessment (cont inued)		
Cert if icat ion Assessment MI LESTONE	Regist r ar per f or ms assessment	
	Pr ocess owner s r esolve assessment f indings	
	Management Rep./ int er nal Assessor s ver if ies closur e of all CAR's	

ITG RESOURCES

IT Governance Ltd sources, creates and delivers products and services to meet the real-world, evolving IT governance needs of today's organisations, directors, managers and practitioners.

The ITG website (*www.itgovernance.co.uk*) is the international one-stop-shop for corporate and IT governance information, advice, guidance, books, tools, training, and consultancy.

www.itgovernance.co.uk/iso9001-quality-management-standards.aspx is the information page on our website for ISO9000 resources.

Other Websites

Books and tools published by IT Governance Publishing (ITGP) are available from all business booksellers and are also immediately available from the following websites:

www.itgovernance.eu is our euro-denominated website which ships from Benelux and has a growing range of books in European languages other than English.

www.itgovernanceusa.com is a US$-based website that delivers the full range of IT Governance products to North America, and ships from within the continental US.

www.itgovernanceasia.com provides a selected range of ITGP products specifically for customers in the Indian sub-continent.

www.itgovernance.asia delivers the full range of ITGP publications, serving countries across Asia Pacific. Shipping from Hong Kong, US dollars, Singapore dollars, Hong Kong dollars, New Zealand dollars, and Thai baht are all accepted through the website.

Toolkits

ITG's unique range of toolkits includes the IT Governance Framework Toolkit, which contains all the tools and guidance that you will need in order to develop and implement an appropriate IT governance framework for your organisation. For a free paper on how to use the proprietary Calder-Moir IT Governance Framework, and for a free trial version of the toolkit, see *www.itgovernance.co.uk/calder_moir.aspx*.

There is also a wide range of toolkits to simplify implementation of management systems, such as an ISO/IEC 27001 ISMS or an ISO/IEC 22301 BCMS, and these can all be viewed and purchased online at *www.itgovernance.co.uk*.

Training Services

IT Governance Ltd offers an extensive portfolio of training courses designed to educate information security, IT governance, risk management, and compliance professionals. Our classroom and online training programmes will help you develop the skills required to deliver best practice and compliance to your organisation. They will also enhance your career by providing you with industry standard certifications and increased peer recognition.

Our range of courses offer a structured learning path from Foundation to Advanced level in the key topics of information security, IT governance, business continuity, and service management.

Full details of all IT Governance training courses can be found at *www.itgovernance.co.uk/training.aspx*.

Professional Services and Consultancy

IT Governance consultants have many years of experience in providing advice, assistance and project support in ISO9001.

Our expert consultants provide you with immediate benefits resulting from the introduction of an ISO9000 Quality Management System. We improve the management of all your internal processes, focusing on the key quality areas that positively affect your ability to win orders and deliver what your customers require, consistently, on time and at a profitable yet economic price.

By using our tried and trusted knowledge transfer approach, we will help you meet the specified requirements that will allow you to achieve certification to the ISO9000 standard and use the quality mark.

ISO9001 does not specify the products, methods, or processes to be used, except in more general terms; the supplier has to identify and document these as part of its own Quality Management System. We can help you to satisfy this requirement in a timely manner.

For more information about IT Governance Consultancy services see:

www.itgovernance.co.uk/iso-9001-quality-management-consultancy.aspx

Publishing Services

IT Governance Publishing (ITGP) is the world's leading IT-GRC publishing imprint that is wholly owned by IT Governance Ltd.

With books and tools covering all IT governance, risk, and compliance frameworks, we are the publisher of choice for authors and distributors alike, producing unique and practical publications of the highest quality, in the latest formats available, which readers will find invaluable.

www.itgovernancepublishing.co.uk is the website dedicated to ITGP enabling both current and future authors, distributors, readers, and other interested parties to have easier access to more information. This allows ITGP website visitors to keep up-to-date with the latest publications and news.

Newsletter

IT governance is one of the hottest topics in business today, not least because it is also the fastest moving.

You can stay up-to-date with the latest developments across the whole spectrum of IT governance subject matter, including risk management, information security, ITIL and IT service management, project governance, compliance, and so much more, by subscribing to ITG's core publications and topic alert emails.

Simply visit our subscription centre and select your preferences: *www.itgovernance.co.uk/newsletter.aspx*.

EU for product safety is Stephen Evans, The Mill Enterprise Hub, Stagreenan, Drogheda, Co. Louth, A92 CD3D, Ireland. (servicecentre@itgovernance.eu)

www.ingramcontent.com/pod-product-compliance
Lightning Source LLC
Chambersburg PA
CBHW061318220326
41599CB00026B/4942